Science

Teacher's Guide

CONTENTS

Author: **Alpha Omega Publications**

Editor: Alan Christopherson, M.S.

Alpha Omega Publications

300 North McKemy Avenue, Chandler, Arizona 85226-2618

O V E R V I E W

SCIENCE

Curriculum Overview
Grades 1–12

Science LIFEPAC Overview

	Grade 1	Grade 2	Grade 3
LIFEPAC 1	YOU LEARN WITH YOUR EYES • Name and group some colors • Name and group some shapes • Name and group some sizes • Help from what you see	THE LIVING AND NONLIVING • What God created • Rock and seed experiment • God-made objects • Man-made objects	YOU GROW AND CHANGE • Air we breathe • Food for the body • Exercise and rest • You are different
LIFEPAC 2	YOU LEARN WITH YOUR EARS • Sounds of nature and people • How sound moves • Sound with your voice • You make music	PLANTS • How are plants alike • Habitats of plants • Growth of plants • What plants need	PLANTS • Plant parts • Plant growth • Seeds and bulbs • Stems and roots
LIFEPAC 3	MORE ABOUT YOUR SENSES • Sense of smell • Sense of taste • Sense of touch • Learning with my senses	ANIMALS • How are animals alike • How are animals different • What animals need • Noah and the ark	ANIMAL GROWTH AND CHANGE • The environment changes • Animals are different • How animals grow • How animals change
LIFEPAC 4	ANIMALS • What animals eat • Animals for food • Animals for work • Pets to care for	YOU • How are people alike • How are you different • Your family • Your health	YOU ARE WHAT YOU EAT • Food helps your body • Junk foods • Food groups • Good health habits
LIFEPAC 5	PLANTS • Big and small plants • Special plants • Plants for food • House plants	PET AND PLANT CARE • Learning about pets • Caring for pets • Learning about plants • Caring for plants	PROPERTIES OF MATTER • Robert Boyle • States of matter • Physical changes • Chemical changes
LIFEPAC 6	GROWING UP HEALTHY • How plants and animals grow • How your body grows • Eating and sleeping • Exercising	YOUR FIVE SENSES • Your eye • You can smell and hear • Your taste • You can feel	SOUNDS AND YOU • Making sounds • Different sounds • How sounds move • How sounds are heard
LIFEPAC 7	GOD'S BEAUTIFUL WORLD • Types of land • Water places • The weather • Seasons	PHYSICAL PROPERTIES • Colors • Shapes • Sizes • How things feel	TIMES AND SEASONS • The earth rotates • The earth revolves • Time changes • Seasons change
LIFEPAC 8	ALL ABOUT ENERGY • God gives energy • We use energy • Ways to make energy • Ways to save energy	OUR NEIGHBORHOOD • Things not living • Things living • Harm to our world • Caring for our world	ROCKS AND THEIR CHANGES • Forming rocks • Changing rocks • Rocks for buildings • Rock collecting
LIFEPAC 9	MACHINES AROUND YOU • Simple levers • Simple wheels • Inclined planes • Using machines	CHANGES IN OUR WORLD • Seasons • Change in plants • God's love never changes • God's Word never changes	HEAT ENERGY • Sources of heat • Heat energy • Moving heat • Benefits and problems of heat
LIFEPAC 10	WONDERFUL WORLD OF SCIENCE • Using your senses • Using your mind • You love yourself • You love the world	LOOKING AT OUR WORLD • Living things • Nonliving things • Caring for our world • Caring for ourselves	PHYSICAL CHANGES • Change in man • Change in plants • Matter and time • Sound and energy

Grade 4	Grade 5	Grade 6	
PLANTS • Plants and living things • Using plants • Parts of plants • The function of plants	**CELLS** • Cell composition • Plant and animal cells • Life of cells • Growth of cells	**PLANT SYSTEMS** • Parts of a plant • Systems of photosynthesis • Transport systems • Regulatory systems	LIFEPAC 1
ANIMALS • Animal structures • Animal behavior • Animal instincts • Man protects animals	**PLANTS: LIFE CYCLES** • Seed producing plants • Spore producing plants • One-celled plants • Classifying plants	**ANIMAL SYSTEMS** • Digestive system • Excretory system • Skeletal system • Diseases	LIFEPAC 2
MAN'S ENVIRONMENT • Resources • Balance in nature • Communities • Conservation and preservation	**ANIMALS: LIFE CYCLES** • Invertebrates • Vertebrates • Classifying animals • Relating function and structure	**PLANT AND ANIMAL BEHAVIOR** • Animal behavior • Plant behavior • Plant-animal interaction • Balance in nature	LIFEPAC 3
MACHINES • Work and energy • Simple machines • Simple machines together • Complex machines	**BALANCE IN NATURE** • Needs of life • Dependence on others • Prairie life • Stewardship of nature	**MOLECULAR GENETICS** • Reproduction • Inheritance • DNA and mutations • Mendel's work	LIFEPAC 4
ELECTRICITY AND MAGNETISM • Electric current • Electric circuits • Magnetic materials • Electricity and magnets	**TRANSFORMATION OF ENERGY** • Work and energy • Heat energy • Chemical energy • Energy sources	**CHEMICAL STRUCTURE** • Nature of matter • Periodic Table • Diagrams of atoms • Acids and bases	LIFEPAC 5
CHANGES IN MATTER • Properties of water • Properties of matter • Molecules and atoms • Elements	**RECORDS IN ROCK: THE FLOOD** • The Biblical account • Before the flood • The flood • After the flood	**LIGHT AND SOUND** • Sound waves • Light waves • The visible spectrum • Colors	LIFEPAC 6
WEATHER • Causes of weather • Forces of weather • Observing weather • Weather instruments	**RECORDS IN ROCK: FOSSILS** • Fossil types • Fossil location • Identifying fossils • Reading fossils	**MOTION AND ITS MEASUREMENT** • Definition of force • Rate of doing work • Laws of motion • Change in motion	LIFEPAC 7
THE SOLAR SYSTEM • Our solar system • The big universe • Sun and planets • Stars and space	**RECORDS IN ROCK: GEOLOGY** • Features of the earth • Rock of the earth • Forces of the earth • Changes in the earth	**SPACESHIP EARTH** • Shape of the earth • Rotation and revolution • Eclipses • The solar system	LIFEPAC 8
THE PLANET EARTH • The atmosphere • The hydrosphere • The lithosphere • Rotation and revolution	**CYCLES IN NATURE** • Properties of matter • Changes in matter • Natural cycles • God's order	**SUN AND OTHER STARS** • The sun • Investigating stars • Common stars • Constellations	LIFEPAC 9
GOD'S CREATION • Earth and solar system • Matter and weather • Using nature • Conservation	**LOOK AHEAD** • Plant and animal life • Balance in nature • Biblical records • Records of rock	**THE EARTH AND THE UNIVERSE** • Plant systems • Animal systems • Physics and chemistry • The earth and stars	LIFEPAC 10

Science LIFEPAC Overview

	Grade 7	Grade 8	Grade 9
LIFEPAC 1	WHAT IS SCIENCE • Tools of a scientist • Methods of a scientist • Work of a scientist • Careers in science	SCIENCE AND SOCIETY • Definition of science • History of science • Science today • Science tomorrow	OUR ATOMIC WORLD • Structure of matter • Radioactivity • Atomic nuclei • Nuclear energy
LIFEPAC 2	PERCEIVING THINGS • History of the metric system • Metric units • Advantages of the metric system • Graphing data	STRUCTURE OF MATTER I • Properties of matter • Chemical properties of matter • Atoms and molecules • Elements, compounds, & mixtures	VOLUME, MASS, AND DENSITY • Measure of matter • Volume • Mass • Density
LIFEPAC 3	EARTH IN SPACE I • Ancient stargazing • Geocentric Theory • Copernicus • Tools of astronomy	STRUCTURE OF MATTER II • Changes in matter • Acids • Bases • Salts	PHYSICAL GEOLOGY • Earth structures • Weathering and erosion • Sedimentation • Earth movements
LIFEPAC 4	EARTH IN SPACE II • Solar energy • Planets of the sun • The moon • Eclipses	HEALTH AND NUTRITION • Foods and digestion • Diet • Nutritional diseases • Hygiene	HISTORICAL GEOLOGY • Sedimentary rock • Fossils • Crustal changes • Measuring time
LIFEPAC 5	THE ATMOSPHERE • Layers of the atmosphere • Solar effects • Natural cycles • Protecting the atmosphere	ENERGY I • Kinetic and potential energy • Other forms of energy • Energy conversions • Entropy	BODY HEALTH I • Microorganisms • Bacterial infections • Viral infections • Other infections
LIFEPAC 6	WEATHER • Elements of weather • Air masses and clouds • Fronts and storms • Weather forecasting	ENERGY II • Magnetism • Current and static electricity • Using electricity • Energy sources	BODY HEALTH II • Body defense mechanisms • Treating disease • Preventing disease • Community health
LIFEPAC 7	CLIMATE • Climate and weather • Worldwide climate • Regional climate • Local climate	MACHINES I • Measuring distance • Force • Laws of Newton • Work	ASTRONOMY • Extent of the universe • Constellations • Telescopes • Space explorations
LIFEPAC 8	HUMAN ANATOMY I • Cell structure and function • Skeletal and muscle systems • Skin • Nervous system	MACHINES II • Friction • Levers • Wheels and axles • Inclined planes	OCEANOGRAPHY • History of oceanography • Research techniques • Geology of the ocean • Properties of the ocean
LIFEPAC 9	HUMAN ANATOMY II • Respiratory system • Circulatory system • Digestive system • Endocrine system	BALANCE IN NATURE • Photosynthesis • Food • Natural cycles • Balance in nature	SCIENCE AND TOMORROW • The land • Waste and ecology • Industry and energy • New frontiers
LIFEPAC 10	CAREERS IN SCIENCE • Scientists at work • Astronomy • Meteorology • Medicine	SCIENCE AND TECHNOLOGY • Basic science • Physical science • Life science • Vocations in science	SCIENTIFIC APPLICATIONS • Measurement • Practical health • Geology and astronomy • Solving problems

Grade 10	Grade 11	Grade 12	
TAXONOMY • History of taxonomy • Binomial nomenclature • Classification • Taxonomy	**INTRODUCTION TO CHEMISTRY** • Metric units and instrumentation • Observation and hypothesizing • Scientific notation • Careers in chemistry	**KINEMATICS** • Scalars and vectors • Length measurement • Acceleration • Fields and models	LIFEPAC 1
BASIS OF LIFE • Elements and molecules • Properties of compounds • Chemical reactions • Organic compounds	**BASIC CHEMICAL UNITS** • Alchemy • Elements • Compounds • Mixtures	**DYNAMICS** • Newton's Laws of Motion • Gravity • Circular motion • Kepler's Laws of Motion	LIFEPAC 2
MICROBIOLOGY • The microscope • Protozoan • Algae • Microorganisms	**GASES AND MOLES** • Kinetic theory • Gas laws • Combined gas law • Moles	**WORK AND ENERGY** • Mechanical energy • Conservation of energy • Power and efficiency • Heat energy	LIFEPAC 3
CELLS • Cell theories • Examination of the cell • Cell design • Cells in organisms	**ATOMIC MODELS** • Historical models • Modern atomic structure • Periodic Law • Nuclear reactions	**WAVES** • Energy transfers • Reflection and refraction of waves • Diffraction and interference • Sound waves	LIFEPAC 4
PLANTS: GREEN FACTORIES • The plant cell • Anatomy of the plant • Growth and function of plants • Plants and people	**CHEMICAL FORMULAS** • Ionic charges • Electronegativity • Chemical bonds • Molecular shape	**LIGHT** • Speed of light • Mirrors • Lenses • Models of light	LIFEPAC 5
HUMAN ANATOMY AND PHYSIOLOGY • Digestive and excretory system • Respiratory and circulatory system • Skeletal and muscular system • Body control systems	**CHEMICAL REACTIONS** • Detecting reactions • Energy changes • Reaction rates • Equilibriums	**STATIC ELECTRICITY** • Nature of charges • Transfer of charges • Electric fields • Electric potential	LIFEPAC 6
INHERITANCE • Gregor Mendel's experiments • Chromosomes and heredity • Molecular genetics • Human genetics	**EQUILIBRIUM SYSTEMS** • Solutions • Solubility equilibriums • Acid-base equilibriums • Redox equilibriums	**CURRENT ELECTRICITY** • Electromotive force • Electron flow • Resistance • Circuits	LIFEPAC 7
CELL DIVISION & REPRODUCTION • Mitosis and meiosis • Asexual reproduction • Sexual reproduction • Plant reproduction	**HYDROCARBONS** • Organic compounds • Carbon atoms • Carbon bonds • Saturated and unsaturated	**MAGNETISM** • Fields • Forces • Electromagnetism • Electron beams	LIFEPAC 8
ECOLOGY & ENERGY • Ecosystems • Communities and habitats • Pollution • Energy	**CARBON CHEMISTRY** • Saturated and unsaturated • Reaction types • Oxygen groups • Nitrogen groups	**ATOMIC AND NUCLEAR PHYSICS** • Electromagnetic radiation • Quantum theory • Nuclear theory • Nuclear reaction	LIFEPAC 9
APPLICATIONS OF BIOLOGY • Principles of experimentation • Principles of reproduction • Principles of life • Principles of ecology	**ATOMS TO HYDROCARBONS** • Atoms and molecules • Chemical bonding • Chemical systems • Organic chemistry	**KINEMATICS TO NUCLEAR PHYSICS** • Mechanics • Wave motion • Electricity • Modern physics	LIFEPAC 10

M
A
N
A
G
E
M
E
N
T

STRUCTURE OF THE LIFEPAC CURRICULUM

The LIFEPAC curriculum is conveniently structured to provide one teacher handbook containing teacher support material with answer keys and ten student worktexts for each subject at grade levels two through twelve. The worktext format of the LIFEPACs allows the student to read the textual information and complete workbook activities all in the same booklet. The easy to follow LIFEPAC numbering system lists the grade as the first number(s) and the last two digits as the number of the series. For example, the Language Arts LIFEPAC at the 6th grade level, 5th book in the series would be LA 605.

Each LIFEPAC is divided into 3 to 5 sections and begins with an introduction or overview of the booklet as well as a series of specific learning objectives to give a purpose to the study of the LIFEPAC. The introduction and objectives are followed by a vocabulary section which may be found at the beginning of each section at the lower levels, at the beginning of the LIFEPAC in the middle grades, or in the glossary at the high school level. Vocabulary words are used to develop word recognition and should not be confused with the spelling words introduced later in the LIFEPAC. The student should learn all vocabulary words before working the LIFEPAC sections to improve comprehension, retention, and reading skills.

Each activity or written assignment has a number for easy identification, such as 1.1. The first number corresponds to the LIFEPAC section and the number to the right of the decimal is the number of the activity.

Teacher checkpoints, which are essential to maintain quality learning, are found at various locations throughout the LIFEPAC. The teacher should check 1) neatness of work and penmanship, 2) quality of understanding (tested with a short oral quiz), 3) thoroughness of answers (complete sentences and paragraphs, correct spelling, etc.), 4) completion of activities (no blank spaces), and 5) accuracy of answers as compared to the answer key (all answers correct).

The self test questions are also number coded for easy reference. For example, 2.015 means that this is the 15th question in the self test of Section II. The first number corresponds to the LIFEPAC section, the zero indicates that it is a self test question, and the number to the right of the zero the question number.

The LIFEPAC test is packaged at the centerfold of each LIFEPAC. It should be removed and put aside before giving the booklet to the student for study.

Answer and test keys have the same numbering system as the LIFEPACs and appear at the back of this handbook. The student may be given access to the answer keys (not the test keys) under teacher supervision so that he can score his own work.

A thorough study of the Curriculum Overview by the teacher before instruction begins is essential to the success of the student. The teacher should become familiar with expected skill mastery and understand how these grade level skills fit into the overall skill development of the curriculum. The teacher should also preview the objectives that appear at the beginning of each LIFEPAC for additional preparation and planning.

TEST SCORING and GRADING

Answer keys and test keys give examples of correct answers. They convey the idea, but the student may use many ways to express a correct answer. The teacher should check for the essence of the answer, not for the exact wording. Many questions are high level and require thinking and creativity on the part of the student. Each answer should be scored based on whether or not the main idea written by the student matches the model example. "Any Order" or "Either Order" in a key indicates that no particular order is necessary to be correct.

Most self tests and LIFEPAC tests at the lower elementary levels are scored at 1 point per answer; however, the upper levels may have a point system awarding 2 to 5 points for various answers or questions. Further, the total test points will vary; they may not always equal 100 points. They may be 78, 85, 100, 105, etc.

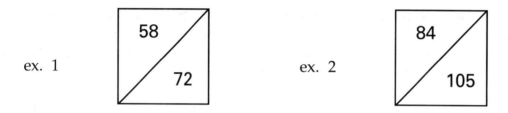

ex. 1   58 / 72   ex. 2   84 / 105

A score box similar to ex.1 above is located at the end of each self test and on the front of the LIFEPAC test. The bottom score, 72, represents the total number of points possible on the test. The upper score, 58, represents the number of points your student will need to receive an 80% or passing grade. If you wish to establish the exact percentage that your student has achieved, find the total points of his correct answers and divide it by the bottom number (in this case 72.) For example, if your student has a point total of 65, divide 65 by 72 for a grade of 90%. Referring to ex. 2, on a test with a total of 105 possible points, the student would have to receive a minimum of 84 correct points for an 80% or passing grade. If your student has received 93 points, simply divide the 93 by 105 for a percentage grade of 86%. Students who receive a score below 80% should review the LIFEPAC and retest using the appropriate Alternate Test found in the Teacher's Guide.

The following is a guideline to assign letter grades for completed LIFEPACs based on a maximum total score of 100 points.

LIFEPAC Test   =   60% of the Total Score (or percent grade)
Self Test   =   25% of the Total Score (average percent of self tests)
Reports   =   10% or 10* points per LIFEPAC
Oral Work   =   5% or 5* points per LIFEPAC
*Determined by the teacher's subjective evaluation of the student's daily work.

Example:

LIFEPAC Test Score	=	92%	92	x	.60	=	55 points
Self Test Average	=	90%	90	x	.25	=	23 points
Reports						=	8 points
Oral Work						=	4 points

TOTAL POINTS	=	90 points

Grade Scale based on point system:	100	–	94	=	A
	93	–	86	=	B
	85	–	77	=	C
	76	–	70	=	D
	Below		70	=	F

TEACHER HINTS and STUDYING TECHNIQUES

LIFEPAC Activities are written to check the level of understanding of the preceding text. The student may look back to the text as necessary to complete these activities; however, a student should never attempt to do the activities without reading (studying) the text first. Self tests and LIFEPAC tests are never open book tests.

Language arts activities (skill integration) often appear within other subject curriculum. The purpose is to give the student an opportunity to test his skill mastery outside of the context in which it was presented.

Writing complete answers (paragraphs) to some questions is an integral part of the LIFEPAC Curriculum in all subjects. This builds communication and organization skills, increases understanding and retention of ideas, and helps enforce good penmanship. Complete sentences should be encouraged for this type of activity. Obviously, single words or phrases do not meet the intent of the activity, since multiple lines are given for the response.

Review is essential to student success. Time invested in review where review is suggested will be time saved in correcting errors later. Self tests, unlike the section activities, are closed book. This procedure helps to identify weaknesses before they become too great to overcome. Certain objectives from self tests are cumulative and test previous sections; therefore, good preparation for a self test must include all material studied up to that testing point.

The following procedure checklist has been found to be successful in developing good study habits in the LIFEPAC curriculum.

1. Read the introduction and Table of Contents.
2. Read the objectives.
3. Recite and study the entire vocabulary (glossary) list.
4. Study each section as follows:
    a. Read the introduction and study the section objectives.
    b. Read all the text for the entire section, but answer none of the activities.
    c. Return to the beginning of the section and memorize each vocabulary word and definition.
    d. Reread the section, complete the activities, check the answers with the answer key, correct all errors, and have the teacher check.
    e. Read the self test but do not answer the questions.
    f. Go to the beginning of the first section and reread the text and answers to the activities up to the self test you have not yet done.
    g. Answer the questions to the self test without looking back.
    h. Have the self test checked by the teacher.
    i. Correct the self test and have the teacher check the corrections.
    j. Repeat steps a–i for each section.

5. Use the SQ3R* method to prepare for the LIFEPAC test.
6. Take the LIFEPAC test as a closed book test.
7. LIFEPAC tests are administered and scored under direct teacher supervision. Students who receive scores below 80% should review the LIFEPAC using the SQ3R* study method and take the Alternate Test located in the Teacher Handbook. The final test grade may be the grade on the Alternate Test or an average of the grades from the original LIFEPAC test and the Alternate Test.

     *SQ3R: Scan the whole LIFEPAC,
             Question yourself on the objectives,
             Read the whole LIFEPAC again,
             Recite through an oral examination, and
             Review weak areas.

GOAL SETTING and SCHEDULES

Each school must develop its own schedule, because no single set of procedures will fit every situation. The following is an example of a daily schedule that includes the five LIFEPAC subjects as well as time slotted for special activities.

Possible Daily Schedule

8:15	–	8:25	Pledges, prayer, songs, devotions, etc.
8:25	–	9:10	Bible
9:10	–	9:55	Language Arts
9:55	–	10:15	Recess (juice break)
10:15	–	11:00	Mathematics
11:00	–	11:45	Social Studies
11:45	–	12:30	Lunch, recess, quiet time
12:30	–	1:15	Science
1:15	–		Drill, remedial work, enrichment*

*Enrichment: Computer time, physical education, field trips, fun reading, games and puzzles, family business, hobbies, resource persons, guests, crafts, creative work, electives, music appreciation, projects.

Basically, two factors need to be considered when assigning work to a student in the LIFEPAC curriculum.

The first is time. An average of 45 minutes should be devoted to each subject, each day. Remember, this is only an average. Because of extenuating circumstances a student may spend only 15 minutes on a subject one day and the next day spend 90 minutes on the same subject.

The second factor is the number of pages to be worked in each subject. A single LIFEPAC is designed to take 3 to 4 weeks to complete. Allowing about 3-4 days for LIFEPAC introduction, review, and tests, the student has approximately 15 days to complete the LIFEPAC pages. Simply take the number of pages in the LIFEPAC, divide it by 15 and you will have the number of pages that must be completed on a daily basis to keep the student on schedule. For example, a LIFEPAC containing 45 pages will require 3 completed pages per day. Again, this is only an average. While working a 45 page LIFEPAC, the student may complete only 1 page the first day if the text has a lot of activities or reports, but go on to complete 5 pages the next day.

Long range planning requires some organization. Because the traditional school year originates in the early fall of one year and continues to late spring of the following year, a calendar should be devised that covers this period of time. Approximate beginning and completion dates can be noted on the calendar as well as special occasions such as holidays, vacations and birthdays. Since each LIFEPAC takes 3-4 weeks or eighteen days to complete, it should take about 180 school days to finish a set of ten LIFEPACs. Starting at the beginning school date, mark off eighteen school days on the calendar and that will become the targeted completion date for the first LIFEPAC. Continue marking the calendar until you have established dates for the remaining nine LIFEPACs making adjustments for previously noted holidays and vacations. If all five subjects are being used, the ten established target dates should be the same for the LIFEPACs in each subject.

FORMS

The sample weekly lesson plan and student grading sheet forms are included in this section as teacher support materials and may be duplicated at the convenience of the teacher.

The student grading sheet is provided for those who desire to follow the suggested guidelines for assignment of letter grades found on page 3 of this section. The student's self test scores should be posted as percentage grades. When the LIFEPAC is completed the teacher should average the self test grades, multiply the average by .25 and post the points in the box marked self test points. The LIFEPAC percentage grade should be multiplied by .60 and posted. Next, the teacher should award and post points for written reports and oral work. A report may be any type of written work assigned to the student whether it is a LIFEPAC or additional learning activity. Oral work includes the student's ability to respond orally to questions which may or may not be related to LIFEPAC activities or any type of oral report assigned by the teacher. The points may then be totaled and a final grade entered along with the date that the LIFEPAC was completed.

The Student Record Book which was specifically designed for use with the Alpha Omega curriculum provides space to record weekly progress for one student over a nine week period as well as a place to post self test and LIFEPAC scores. The Student Record Books are available through the current Alpha Omega catalog; however, unlike the enclosed forms these books are not for duplication and should be purchased in sets of four to cover a full academic year.

WEEKLY LESSON PLANNER

Week of:

	Subject	Subject	Subject	Subject
Monday				
	Subject	Subject	Subject	Subject
Tuesday				
	Subject	Subject	Subject	Subject
Wednesday				
	Subject	Subject	Subject	Subject
Thursday				
	Subject	Subject	Subject	Subject
Friday				

WEEKLY LESSON PLANNER

Week of:

	Subject	Subject	Subject	Subject
Monday				
	Subject	Subject	Subject	Subject
Tuesday				
	Subject	Subject	Subject	Subject
Wednesday				
	Subject	Subject	Subject	Subject
Thursday				
	Subject	Subject	Subject	Subject
Friday				

Student Name _____   Year _____

Bible

LP #	Self Test Scores by Sections 1	2	3	4	5	Self Test Points	LIFEPAC Test	Oral Points	Report Points	Final Grade	Date
01											
02											
03											
04											
05											
06											
07											
08											
09											
10											

History & Geography

LP #	Self Test Scores by Sections 1	2	3	4	5	Self Test Points	LIFEPAC Test	Oral Points	Report Points	Final Grade	Date
01											
02											
03											
04											
05											
06											
07											
08											
09											
10											

Language Arts

LP #	Self Test Scores by Sections 1	2	3	4	5	Self Test Points.	LIFEPAC Test	Oral Points	Report Points	Final Grade	Date
01											
02											
03											
04											
05											
06											
07											
08											
09											
10											

Student Name _____   Year _____

Mathematics

LP #	Self Test Scores by Sections 1	2	3	4	5	Self Test Points.	LIFEPAC Test	Oral Points	Report Points	Final Grade	Date
01											
02											
03											
04											
05											
06											
07											
08											
09											
10											

Science

LP #	Self Test Scores by Sections 1	2	3	4	5	Self Test Points	LIFEPAC Test	Oral Points	Report Points	Final Grade	Date
01											
02											
03											
04											
05											
06											
07											
08											
09											
10											

Spelling/Electives

LP #	Self Test Scores by Sections 1	2	3	4	5	Self Test Points	LIFEPAC Test	Oral Points	Report Points	Final Grade	Date
01											
02											
03											
04											
05											
06											
07											
08											
09											
10											

N
O
T
E
S

INSTRUCTIONS FOR SCIENCE

The LIFEPAC curriculum from grades two through twelve is structured so that the daily instructional material is written directly into the LIFEPACs. The student is encouraged to read and follow this instructional material in order to develop independent study habits. The teacher should introduce the LIFEPAC to the student, set a required completion schedule, complete teacher checks, be available for questions regarding both content and procedures, administer and grade tests, and develop additional learning activities as desired. Teachers working with several students may schedule their time so that students are assigned to a quiet work activity when it is necessary to spend instructional time with one particular student.

The Teacher Notes section of the Teacher's Guide lists the required or suggested materials for the LIFEPACs and provides additional learning activities for the students. The materials section refers only to LIFEPAC materials and does not include materials which may be needed for the additional activities. Additional learning activities provide a change from the daily school routine, encourage the student's interest in learning and may be used as a reward for good study habits.

If you have limited facilities and are not able to perform all the experiments contained in the LIFEPAC curriculum, the Science Project List for grades 3-12 may be a useful tool for you. This list prioritizes experiments into three categories: those essential to perform, those which should be performed as time and facilities permit, and those not essential for mastery of LIFEPACs. Of course, for complete understanding of concepts and student participation in the curriculum, all experiments should be performed whenever practical. Materials for the experiments are shown in Teacher Notes – Materials Needed.

Science Projects List

Key

(1) = Those essential to perform for basic understanding of scientific principles.

(2) = Those which should be performed as time permits.

(3) = Those not essential for mastery of LIFEPACs.

S = Equipment needed for home school or Christian school lab.

E = Explanation or demonstration by instructor may replace student of class lab work.

H = Suitable for homework or for home school students. (No lab equipment needed.)

Science 301

pp	6	(1)	H
	11	(2)	H
	23	(1)	H
	30	(1)	H
	31	(1)	H or S

Science 302

pp	6	(1)	H or S
	8	(1)	H
	16	(3)	E
	22	(3)	H
	27	(2)	H
	30	(1)	H

Science 303

pp	7	(1)	S
	9	(2)	S
	11	(2)	S
	17	(3)	S
	19	(3)	H
	41	(1)	S
	49	(2)	S

Science 304

pp	37	(3)	S

Science 305

pp	6	(3)	H
	7	(3)	H
	13	(3)	H
	17	(2)	S
	18	(1)	E
	20	(1)	S
	24	(1)	S
	28	(2)	H
	35	(1)	S

Science 306

pp	5	(1)	S
	7a	(1)	S
	7b	(1)	S
	16	(2)	H

19	(2)	S
20	(3)	H
22	(2)	S
28	(3)	H
31	(2)	S

Science 307

pp	5	(2)	S
	11	(3)	H
	12	(1)	H
	27-31	(1)	S
	40	(2)	H

Science 308

pp	19	(1)	S
	21	(2)	S
	24	(1)	H
	52	(3)	H
	60	(3)	H

Science 309

pp	7	(2)	H
	12	(1)	S
	15	(1)	H
	17	(2)	H
	21	(1)	H
	24	(3)	S
	33	(2)	H
	35	(1)	H
	39	(1)	H
	41	(1)	H
	46	(1)	E

Science 310

pp	31	(1)	H
	37	(2)	H
	41	(1)	S

Materials Needed for LIFEPAC:

Required:
clock with a second hand
salt
cocoa
paper towel
small drinking glass
large drinking glass half full of water

Suggested:
sponge

Additional Learning Activities

Section I *Your Body Breathes Air*

1. If a model of the lungs is available, use it for discussion. If not available, use pictures. Discuss placement and function of lungs, and so forth.
2. To show that air is everywhere, put soil into a jar. Slowly pour water into the jar (bubbles should rise). Ask students to record what they observe.
3. Ask students to blow up a balloon. Release the air slowly. Ask them to record what changes they see in the balloon both before and after it has been inflated. Write a report comparing the balloon to the lungs.
4. Make a picture showing the "balance of nature." People and animals need oxygen. Plants need carbon dioxide. Both should be included in the picture (may be colored or painted).

Section II *Your Body Digests Food*

1. Cut very small pieces of a turnip, potato, carrot, zucchini, and so forth for a "tasting" party. Ask students to describe taste. How can they tell what vegetable they are eating?
2. Have students put a teaspoon of salt in water. Then put a piece of apple in the water. Record what happens to the salt and to the apple.
3. On a sheet of paper, write the process of digestion:
   a. teeth and saliva,
   b. the stomach,
   c. the small intestine,
   d. in the blood, and
   e. large intestine stores food that is not used and it passes from your body.

Section III *Your Body Exercises and Rests* .

1. Play "Jump the Shot." Make a circle. Drop hands. One person stands in the middle. He has a jumping rope with a bean bag tied tightly at the other end. He swings the rope around close to the ground. As the rope goes around, others jump. Change the rope swinger frequently. Stop occasionally to feel how much faster breathing and heartbeat is.
2. Have the student make a poster on 12" x 18" tagboard to show what can be done to keep muscles and bones healthy (exercise, rest, good food, good posture, etc.).

Section IV *Your Body is Different from an Animal's*

1. Make a list on the board of things that people need to grow: food, sleep, exercise, good health. Discuss.

2. Make a list of things that show how you can tell boys and girls are "growing up." For example:
   a. getting bigger,
   b. using tools better,
   c. working more (cleaning your room),
   d. thinking of others,
   e. being more responsible,
   f. keeping clean, and
   g. being creative.
   Ask students to choose one of these items and make a picture of it.
3. Have students make a picture showing three or four boys and girls in the classroom. Show that not everyone grows at the same rate.
4. Ask students to write a short story about growth from birth to the present time. Students might use pictures to go with the story.
5. Ask students to think of an area where they are most creative. Then have them write about it or create something in class.

Materials Needed for LIFEPAC

Required:
small, living plants and a magnifying glass
small jar with water
food coloring
celery stick
magnifying glass
glass jar
wet paper towel
three lima beans
felt marker
a centimeter ruler

Suggested:
two small potted plants, box, and a sunny
  window
jar with water
fifteen lima beans
potting soil
five paper cups
water
white potato (with "eyes")
knife

Additional Learning Activities

Section I *Plant Parts*

1. Have students make lists of two things roots do, one thing stems do, and one thing leaves do.
2. Have committees of students draw a picture of plants. Show all parts both above and below the ground. Use large sheets of banner paper.
3. If possible, plan a field trip to an orchard, large university garden, or a botanical garden.
4. Have students make a poster on drawing paper, or on tagboard, showing the kinds of plants we eat for food. Try to draw some stems, roots, flowers or seeds.
5. Have students draw three columns on a piece of writing paper. In row one, make a list of plants of which we use the roots for food. In the second, list the plants of which we use the stems for food, and in the third, make a list of plants of which we use the leaves for food. Students may need to use an encyclopedia or other reference material.
6. Visit a supermarket. Take a notebook. Go to the fresh produce section and make a list of all the vegetables and fruits that you see. Divide them into four groups in your notebook. List each item under one of the following headings: (a) food stored in roots, (b) food stored in stems, (c) food stored in seeds, and (d) food stored in fruit.

Section II *Plant Growth*

1. Find a healthy spot of grass outdoors. Place a board over part of it. Leave the board several days. Remove the board. Discuss why the grass is no longer green.
2. If it is extremely hot or cold outdoors, place one plant out and keep a similar plant indoors. Report orally or on paper what happened to both plants and why.
3. Ask the student to write the name of some plants he saw on the way to school. Write a report about some of these plants.
4. On a piece of paper, ask the students to choose one plant and show how it gets its necessary requirements. Tell them to use their imagination in deciding what the plant might look like underground (where it gets its water, etc.).

Section III *Plant Changes*

1. Display pictures of plants. Discuss the parts and what they do for the plant.
2. Ask students to tell why food is important to them. Cut pictures from magazines or seed catalogs to show what foods are important to them. Paste these on a large piece of banner paper. (All students put their pictures on the same mural.)
3. Have students make a seed collection. Use poster board. Glue seeds on poster board with white glue. Label each kind of seed carefully. Students will need to collect their seeds; therefore, time will need to be considered.
4. Make pictures of a plant from seed or bulb to a fully grown plant with seeds. Number each picture to show how it grows. Students may choose any plant they wish.

Example: In the first picture, draw a picture of an acorn; next, a sprout; and then, a seedling. Finally, you may show an oak tree at different heights. The last picture will be a tall oak tree with acorns on it.

Materials Needed for LIFEPAC

Required:
clock
Celsius thermometer
Fahrenheit thermometer
gallon jar
a lid with some holes in it that will fit the jar
moist soil
potato
3 or 4 sow bugs (sometimes
called potato bugs)

Suggested:
two-pint glass jars or bowls
2 thermometers
soil, water, and ruler
quart jar
moist soil
tray
2 or 4 earthworms
watch with second hand
a cup
2 or 4 jars with lids
water

Additional Learning Activities

Section I *What Changes an Environment?*

1. Show students pictures of different animals found near ponds, in forests, and in the deserts. Ask them to identify each animal and describe its activities as shown in the picture. Conduct discussion related to the student's own experiences in observing similar animals (pets, zoo animals, etc.).
2. Write "Words to Study" on chalkboard in mixed order. Have students tell what each word means and have individuals go to chalkboard and erase words as they get them correct. Drill on words as needed.
3. Have students find pictures of animals that live near their homes. Have them select the best pictures for showing need of food, water, and sunlight. Display pictures in classroom.
4. Two or more students may play a game using 3" x 5" cards with "Words to Study" printed on them. One student holds up card; another student gives the meaning of the word and secures the card. The student who gets the most cards wins after repeating the game with the roles reversed.
5. Student may select a good library book about the life of some animal and read an interesting portion to the class.
6. Individual student may wish to collect samples of materials which make up the earth. Different kinds of soil and water from lakes, rivers, and oceans might be labeled and displayed.

Section II *How Are Animals Different?*

1. Plan a class visit to a natural history museum or zoo. Have the class take notes and report on structure, food, and respiration as related to animals.
2. Discuss with students how mouths of frogs, woodpeckers, tigers, squirrels, and ducks are adapted for getting food. Use pictures from local libraries or your personal picture file.
3. Have a small group of interested students prepare a bulletin board display entitled "How Do Animals Get Their Food?"
4. Have the group collect pictures of scuba divers, deep-sea divers, and astronauts. Then discuss why these persons need special equipment for breathing. Display the best and explain the different parts of it.

5. Have a student tell the class about a fishing trip he has taken or read about. Tell about the kinds of fish that can be most likely caught by using worms and the kinds that are most likely caught by using fish as bait.

6. Place a snail in a bowl or tank of water and observe and report on its ability to move about in the tank.

7. Collect pictures of animals from old magazines. (*Nature Magazine, National Geographic,* etc.) Organize pictures into groups, such as animals that eat plants, animals that eat animals, and animals that eat both plants and animals. Display at open house or in the classroom for students to study.

Section III *How Do Animals Grow and Change?*

1. Find pictures of baby penguins and compare them with pictures of the adult bird. Find a story of how parent penguins care for their young.

2. Visit a chicken hatchery with children or hatch chickens in the classroom.

3. Have the students set up a terrarium. Bring a praying mantis and a frog or toad and place them in the terrarium. Students should make careful observations and report how these animals lie in wait to catch insects.

4. Students may bring a fish or frog into the classroom and a small group may observe it for a few days. One member of the group may write a report; another member could give an oral report. Students skilled in art may illustrate the written reports or display their art work to the class.

5. Prepare a special report on one of the following animals: seals, lemmings, or prairie dogs.

6. Students may show pictures of farm animals, zoo animals, or pets. They should compare the young animal with its parents.

Materials Needed for LIFEPAC

Required:
numerous magazines with pictures of food

Suggested:
restaurant menus
large plate
fruit that can be easily cut up
toothpicks
napkins
cream cheese
cheese
celery
peanut butter

Additional Learning Activities

Section I *How to Build a Healthy Body*

1. Direct a discussion on additives.
2. Have the student draw, cut, or paste pictures on a paper plate showing a balanced meal. Place small wall hanger on back for display in the classroom. Decorate edge of plate.
3. Make a food dictionary for open house.

Section II *How To Plan For Healthy Eating*

1. Have several students make a food dictionary and take a trip to the local supermarket. Check healthful foods and identify them on the students' lists with an X for healthful foods and an O for "junk" foods.
2. The group may collect pictures of table settings and food from household magazines. Hold a small-group discussion on the topic "How Advertising (magazine, television, and billboard) Is Influencing Our Eating Habits for Good or Bad."
3. With parents' permission a student may prepare a balanced meal at home and write or give an oral report on "My Experience in Making a Balanced Meal."
4. Do special research on scurvy and report in class.

Section III *How To Form Good Health Habits*

1. The group may make a bulletin board display on care of the teeth.
2. A small group may prepare a table-top collection and a display of different kinds of toothpaste, dental floss, and pictures of tooth care.
3. The student may do research and make an individual report on "junk" foods.
4. The student may give a demonstration on the proper cleaning of one's teeth.
5. The student may report on a visit to the dentist, telling about any important advice the dentist gave him.

Materials Needed for LIFEPAC

Required:
three balloons
wire coat hanger
door
food scale that reads in grams
baby-food jar
5 or 6 ice cubes
a soft drink bottle
vinegar
baking soda
a cone-shaped drinking cup
a measuring cup
a plastic teaspoon

Suggested:
pan of water
cup of sand
stick
sandy water from previous experiment
bowl
strainer either metal or cloth
3 different sizes of books
3 different sizes of bottles filled with water
3 different sizes of rocks
a card with *small* printed on it
a card with *middle* printed on it
card with *big* printed on it
brick
5 paper cups
5 ice cubes
cool water
hot water
clock

Additional Learning Activities

Section I *Matter*

1. Have a student prepare a special report on Robert Boyle.
2. A student may make a table display of common objects found in the classroom. Label each article with a list of at least three properties of that object (see Section I of LIFEPAC).

Section II *Change in Matter*

1. A group of students may make a "magic garden." Make a hot sugar solution. Add food coloring to it. Place some porous material such as charcoal in a pie pan. Pour the colored solution over it. It will soak upward and as the water evaporates, crystals will form. Have students discuss evaporation and crystals among themselves and, after reading more about this topic, report to the class.
2. Have the group make lists of as many physical changes as they can. The one who has the longest correct list is considered the winner. Allow a specified time. The winner may wish to read his list to the class.

Materials Needed for LIFEPAC

Required:
drum
dry cereal
pencil
tuning fork

Suggested:
2 pieces of rubber to fit over ends of box
an oatmeal box
a hole punch
heavy string
scissors
pencil
ruler
water in large glass
stones
small, strong box open on one side
rubber bands of different sizes
large pan
small pan
ruler
string
the drum previously made
clock that ticks, a roll of paper, and a
   yardstick
balloon, a rubber band, a can with both
   ends removed
friend
dry cereal

Additional Learning Activities

Section I *Sounds Are Made*

1. Have students remain very still for two minutes. Tell them you will ask them to write down all the sounds they heard during the two minutes. Have them share what they heard.
2. Have students close their eyes. Strike or blow several different instruments (toys, rhythm band instruments, piano, etc.) and ask the students to guess what instrument it is.
3. Have several students find books about sound in the local library and share what they have found with each other.
4. Show pictures from your picture file of numerous musical instruments. Discuss what causes each instrument to make a sound.
5. Have a student explain how a tuning fork makes sound.
6. Have a student research and write a short report on how radar works.
7. Obtain one or more of the following items: vase, fruit jar, milk bottle, small pitcher, or large sea shell. Put each of them up to your ear and listen. What do you hear? How does this experiment show that sea shells do not hold the sound of the sea? Explain to the class what causes these objects to make a slight roaring noise.

Section II *Sounds Are Different*

1. Blindfold a friend (at recess or noon-hour). Move to different parts of the room. Find out how well your friend can tell where you are without seeing you. Are ears good for telling direction?
2. Have students work in pairs to list sounds from some of the following places: church, street corner, bus, schoolyard, supermarket, airport, or shop. Report to class.
3. Have students bring in items that make unusual sounds.
4. Student may make a musical instrument from five water glasses. Fill each glass with water to a different height. Line the glasses up and tap gently with a pencil. Experiment with water level until student can play a tune on them.

Section III *Sounds Are Heard*

1. Two or three students may work together to build a simple telegraph sender and receiver. Students might study Morse code and send simple messages or learn the Morse code alphabet.
2. Have several students bring musical instruments to class and experiment making different sounds. Each student should report on how he made his instrument vibrate.
3. Listen to sounds through a hollow tube. Then stuff paper into the tube. Does sound travel well through the paper? Listen to sounds through other materials. Compare and report.
4. The student may secure directions from a library book and make a tin-can telephone. Have him explain to the class how he did it and have a friend talk with him on the tin-can telephone.
5. The student may build a model of the ear out of papier-maché.

Materials Needed for LIFEPAC

Required:
clay
3 pins
pencil
flashlight
globe
chalk

Suggested:
bright flashlight
glove
paper-towel tube
tape
lamp
globe with your town marked on it

Additional Learning Activities

Section I How the Earth Moves

1. Have the group collect clocks and set them for the different time zones in the world.
2. Make up and play a game called "What Time Is It?" Write questions on 3" x 5" cards, such as "When it is one o'clock in (your home town), what time is it in ?"
3. Make a map of the time zones in the United States. Show, by drawings or pictures from old magazines, what someone is doing in each time zone. (Some telephone directories have time zone maps.)
4. Consult an encyclopedia or almanac and draw and color the time zones of the world.

Section II Why Time Changes

1. Make a simple sundial. Fasten a stick standing straight up on a board. Place the board in the sun. Use a watch and record where the end of the shadow of the stick is at each hour. After the hours are marked on the board, explain how you can tell time by using a sundial.
2. Using an almanac, keep a record of the time the sun rises and sets for a week or two. Keep a record of the hours between sunrise and sunset during this time. At what conclusions do you arrive?
3. Research the history of the sundial and make a report to the class on the subject.

Section III Why Seasons Change

1. Four students may join together in writing stories about the seasons. Each student should take one season. The completed stories (when checked by the teacher) can then be placed in a folder entitled "Stories About the Seasons." Another student may wish to illustrate the book.
2. Make a class scrapbook of each season using pictures from old magazines.
3. Make a mural of the seasons in their order.
4. Write a play about the seasons.
5. Make some star maps.

Materials Needed for LIFEPAC

Required:
piece of limestone
pocket knife
some vinegar
spoon
glass or plastic jar
string
magnifying glass
table salt
small, flat stick
tablespoon
some warm water

Suggested:
rock collection with labels
glass or plastic jar with cover
coarse sand
small pebbles
water
soil

Additional Learning Activities

Section I *How Rocks Are Formed*

1. Have students look at objects around the classroom, at home, and around the schoolyard. Identify and make a list of items that contain minerals. See who can obtain the longest list over a two-day period. (Minerals to look for may include silver, gold, copper, diamond, sulfur, oil, lead, iron, zinc, and talc.)
2. Write a short radio script about some topic of interest connected with this LIFEPAC. Simulate a radio broadcast using a microphone and sound effects. Members of the group read script.
3. Draw pictures to show how igneous and sedimentary rocks are formed.
4. Write and read an imaginary radio announcement about a volcanic eruption somewhere in the world. Write from the viewpoint of a reporter flying near the volcano.

Section II

1. Show pictures of an erupting volcano (*National Geographic* magazines, *Travel* magazines, etc.) and discuss with class. Discuss superstitions regarding volcanoes and why ancient people feared them (for instance, Pele in Hawaii).
2. Each member of a group may choose a state and find out what minerals and rock products are listed as important from that state. Refer to encyclopedia or other reference books, and compare lists. One member of the group may wish to compile into one longer list.
3. Students may read about lunar rocks. How are they different from earth rocks?
4. A student may do individual reading about "Rock City" near Minneapolis, Kansas, and tell the class about this unusual rock formation.

Section III *How Rocks Are Used*

1. Several students may work together to make a list of gems. Write one statement about each gem. Use several reference books for research.
2. Make a special study of cement. Compare the uses of cement with stone and rock for building purposes. Each member give a report to the class on a particular segment of the study.

3. Collect small pieces of plaster, brick, and concrete to see how they compare to actual rock. Plan an exhibit for the science table (or for open house). Concrete is similar to conglomerate. Include some good books on the subject in the exhibit.
4. Make a table display of equipment needed to go rock hunting (hammer, chisel, goggles, magnifier, compass, etc.).
5. Interested students may compile a list of superstitions about rocks. Explain why we do not believe in superstition today.
6. Write a "Letter to the Editor" about why rock collectors should not go hunting rocks on private property without permission from the owner.

Materials Needed for LIFEPAC

Required:	Suggested:
3 candles in holders	wood
pint jar	a nail
quart jar	sandpaper
clock or timer	hammer
matches	mirror
small, flat box	lined paper
plastic food wrap	hairbrush
small bits of tissue paper	dark room or closet
small-necked bottle	comb
pan of water	sunny day
balloon	magnifying glass
hot plate	a piece of paper
large metal can	2 clear glasses
pan or kettle	food coloring
one-cup measure	hot water
glass of water	dropper
pot holder	cold water
ruler	a clock or timer
six thumbtacks	
copper tube	
wooden stick	

Additional Learning Activities

Section I *Where Heat Energy Comes From*

1. Have the class (or small group studying heat energy) visit a solar-heated home to learn how sun power works.
2. Make a solar cooker. Make a cone out of tinfoil and face the inside toward the sun. Put pieces of hot dogs in the center and watch them cook.
3. Find out how hot-air balloons work. Read about them in a magazine and draw or paint some pictures to put in a class scrapbook.
4. Student may "Ask an Expert" (see accompanying Worksheet Number 1).

Section II *What Heat Energy Is and Does*

1. Student may experiment with ice. Independent activity may be duplicated (see Worksheet Number 2).
2. Group collects clippings of new experiments in sources of energy. Keep these in a class scrapbook. Add original drawings or pasted magazine pictures.
3. Hold a magnifying glass to a pinwheel. What happens? That is solar power.
4. Make drawings showing what heat does to a solid, to a gas, and to a liquid.

Section III *How Heat Energy Affects Our Lives*

1. Discuss the meaning of *conservation* and *scarcity*. Take an apple and ask the students what they must do if they want to have something left of the apple

tomorrow. If people in the classroom all want to share this apple, what must you do? If this is the only apple left in the whole world and you are hungry, what will you do then?

2. Organize a tabletop science display about energy.

3. Students may visit a local power plant to learn how electricity is produced. Discuss pollution control.

4. Student with camera may take snapshots of pollution in streams and ponds. Have group discuss problem *of* pollution.

5. Student may find out how to read an electric meter. Read the meter at home a week or two apart and report on number of kilowatts used. Tell how we can save electricity.

Materials Needed for LIFEPAC

Required:
2 cardboard boxes (same size)
black paint
white paint
brush
2 thermometers

Suggested:
string
wire
chair
coat hanger

Additional Learning Activities

Section I *Physical Change*

1. Grow seeds indoors.
   a. Place the seeds in a cup of tea.
   b. Place the cup in the refrigerator.
   c. Leave in the refrigerator four days.
   d. At the end of four days, remove them. Drain off the tea and spread the seeds out to dry.
   e. When dry, plant the seeds in sphagnum moss in a metal or plastic container.
   f. Keep a record of changes and growth.
2. Do research to discover why seeds grow better if they are first soaked in tea and left in the refrigerator for four days.

Section II *The Earth*

1. Try to obtain some silkworm eggs
   a. Keep eggs in a warm place for two weeks.
   b. When the silkworms hatch, add mulberry leaves. Put a plastic covering over the container to prevent drying out.
   c. Add fresh mulberry leaves and remove debris until worms are four weeks old.
   d. Cut a carton two inches high, with a bottom panel for putting in new leaves. Cover top and bottom of carton with saran wrap to waterproof and allow for viewing.
   e. Allow one week for silkworms to spin their cocoons.
   f. Allow another two weeks for the moths to emerge from their cocoons. (NOTE: This activity should be started early in the school year to allow for completion.)
2. Students assist with setting up silk spinning activity.
3. Students observe growth and changes in silkworms at various stages of metamorphosis. Cocoon stage lasts about two or three weeks.
4. After moths have emerged from the cocoons, boil the cocoons in water to loosen the silk. Student should wind the silk threads on a spool and see and feel the material.
5. Student may wish to draw the various stages of development (egg, larva, pupa, adult moth) as they occur. (NOTE: If this activity is not practical in a particular classroom, several of the activities may be carried out by using reference books only.)

6. Student may bring a plant (or weed) from home. Have him make a drawing of his plant and label the basic parts.

7. Student may bring pictures of different animals cut from magazines and newspapers from home. Student may use pictures to create a bulletin board showing both vertebrates and invertebrates.

8. This test can be done to find out if a rock has lime in it. Find several rocks. Get a bottle of vinegar and pour a little over each rock. If the rock bubbles up it contains lime and is limestone.

9. **Food Booklets:** Child staples blank sheets of paper together and adds fancy wallpaper covers. The booklets can be used as dictionaries, by writing food words on the right page (apple on first page, zucchini on last, etc.). You can organize a cookbook of children's favorite recipes or meals and put them in the booklets. Also you can make up food riddles and write in your food booklet (e.g.: I am white. I am in the vegetable group. You eat my flowers. I taste a little bit like cabbage. Who am I? Answer: cauliflower).

10. Solving the Mystery of the Seed Set up a demonstration experiment involving bean seeds to show the growth and development of roots, stems, and leaves. With the correct water and temperature, a young plant will grow from each seed. It uses food in the seed to grow.

    Procedure:
    a. Soak lima beans overnight.
    b. Line a glass container with a paper towel or cardboard like paper.
    c. Now place each seed between the paper and the glass.
    d. Remove one of the seed leaves from one of the seeds and place it between the glass and the paper.
    e. Take another seed and cut it in half and place it between the glass and the paper.
    f. Keep water in the bottom of the glass.
    g. Watch them grow.

Notes

TESTS & KEYS

Reproducible Tests
for use with the Science 300
Teacher's Guide

Name _____

Draw a line to match the words with the right phrase.

1. heart   pump blood

2. conscience   openings in nose

3. saliva   man has

4. stomach   digestive system

5. nostrils   breathe out

6. exhale   in mouth

Number the places that food goes in the order of the digestive system.

7. _____ small intestine

8. _____ stomach

9. _____ food tube

0. _____ mouth

1. _____ large intestine

Fill in the circle in front of the right answer.

2. The scientist who discovered oxygen was _____.

    O William Jacobs   O Jacob Pride   O Joseph Priestly

3. The breathing muscle is the _____.

    O stomach   O head   O diaphragm

4. When the heart beats slower, the blood moves _____.

    O slower   O faster   O at the same rate

15. The body inhales _____.

    O carbon dioxide   O food   O oxygen

16. Air in the nose is cleaned by _____.

    O pulse   O hairs   O food

Complete these activities.

17. Name two ways people and animals are the same.

    a._____

    b._____

18. Name three ways people are different from animals.

    a._____

    b._____

    c._____

Fill in the circle in front of the right answer.

19. To know how fast the heart beats you feel the _____.

    O bones   O nose   O pulse

20. Blood takes oxygen to all parts of the _____.

    O body   O toes   O fingers

21. To build a strong body, you must _____.

    O eat too much   O exercise   O relax

22. When you take air into your lungs, you _____.

    O inhale   O exhale   O blow

23. Animals do *not* have a _____.

    0 body   0 head   0 spirit

24. Animals do *not* have _____.

    0 toes   0 a conscience 0 a pair of lungs

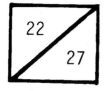

Date _____

Score _____

Name _____

Draw lines to match the word with the right phrase.

1. roots   used to make things look larger

2. oxygen
                           take in water and minerals

3. magnifying glass
                           take water and minerals to the leaves

4. leaves
                           where some plants store food

5. stems
                           gas given off by plants

Write the answers from the list on the lines.

food   sprout   temperature

roots   carbon dioxide   chlorophyll

seed

6. The green coloring material in plants is _____.

7. A tiny plant and its food make a _____.

8. A gas that plants take from the air is _____.

9. God gave us green plants for _____.

10. Some plants store food in the _____.

11. A plant must have the right _____ in order to grow.

12. When a plant just begins to grow, it is called a _____.

Draw a line from the plant part to the name of a part that is eaten.

13. leaves   carrots

14. fruit   beans

15. seed   celery

16. stem   lettuce

17. roots   apple

Circle the right answer.

18. Plant roots _____ .

    a. take in chlorophyll

    b. take in water and minerals

    c. store sunlight

19. Plant leaves _____ .

    a. give off oxygen and water

    b. store water

    c. make seeds

20. Plant stems _____ .

    a. give shade

    b. give off carbon dioxide

    c. are like tiny pipes

Complete this activity.

21. Write five things a plant needs to grow.

a. _____

b. _____

c. _____

d. _____

e. _____

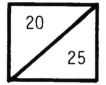

Date _____

Score _____

Name _____

Fill in the circle in front of the right answer.

1. The boiling point of water is _____.

   O 32° F   O 212° F   O 0° C

2. The melting point of ice is _____.

   O 100° C   O 212° F   O 0° C

3. The freezing point of water is _____.

   O 32° F   O 212° F   O 100° C

4. An instrument used to measure temperature is a _____.

   O thermometer   O vertebrate   O bird

5. Soil heats _____ water

   O slower than   O faster than   O at the same rate as

6. Animals that have backbones are called _____.

   O invertebrates   O bugs   O vertebrates

7. Sowbugs are also called _____.

   O birds   O potato bugs   O water bugs

8. Ice floats because water _____ when it freezes.

   O expands   O gets heavier   O melts

9. The name of the change that some invertebrates go through is called _____.

   O metamorphosis   O mammals   O vertebrates

10. Animals that give birth to their young and make milk for them are _____.

    O water animals   O land animals   O mammals

Complete the following lists.

11. List four things that can change environments.

    a._____   c._____

    b._____   d._____

12. List the five groups of vertebrates.

    a._____   d._____

    b._____   e._____

    c._____

13. List the four stages in the metamorphosis of a butterfly.

    a._____   c._____

    b._____   d._____

| 18 / 23 |

Date _____

Score _____

Name _____

Write the correct word from the list on the line. Some words may be used more than once.

energy   clothes   grow

shoulder   sick   water

God   junk   clean

1. You should drink four to six glasses of _____ each day.

2. The One who wants you to keep your thoughts clean is _____.

3. Food gives you _____.

4. Food helps you to _____ taller.

5. Soft drinks are _____ food.

6. You should read with a light shining over your _____.

7. You should brush your teeth the way they _____.

8. You should wear clean _____ every day.

9. God wants you to keep your thoughts _____.

10. Good food can help keep you from getting _____.

Do this activity.

11. Write the six food groups.

a. _____

b. _____

c. _____

d. _____

e. _____

f. _____

Answer *true* or *false*.

12. _____ Food helps you keep warm.

13. _____ Food helps you obey your father.

14. _____ Food helps to keep you healthy.

15. _____ Food helps you to answer the phone.

16. _____ Milk is a good food.

17. _____ Food helps you gain weight and grow.

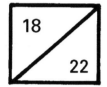

Date _____

Score _____

Name _____

Draw a line to match.

1. **Melting ice**   fixed shape and size

2. God   chemical change

3. gas   physical change

4. forest fire   cannot be seen

5. atom and molecules   made all matter

6. invisible   changes shape and size

7. solid   tiny bits of matter

8. volume   how much **matter** is in something

9. mass   a person that studies chemistry

10. chemist   space used by matter

Fill in the circle in front of the right answer.

11. **Baking bread is a** _____.

    O chemical change   O physical change

12. **Freezing water into ice is a** _____.

    O chemical change   O physical change

13. When a liquid becomes cold, it becomes a _____.

    O gas   O solid

Write *yes* or *no*.

14. _____ Melting ice is a physical change.

15. _____ Some matter has no volume.

16. _____ All things are made of matter.

17. _____ All matter has mass.

18. _____ Liquid has size and changing shape.

19. _____ Melting ice cream is a physical change.

20. _____ A puddle of water drying up is a chemical change.

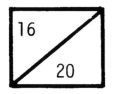

Date _____

Score _____

Name _____

Write *yes* or *no*.

1. _____ You cannot hear sounds that come from above you.

2. _____ Sounds do not travel.

3. _____ If a person's hearing is poor, nothing can be done to help it.

4. _____ Sounds are vibrations.

5. _____ All sounds are the same.

6. _____ The eardrum vibrates when sound goes into the ear.

7. _____ Your brain takes messages that the nerves bring.

8. _____ Some sounds are made without vibrations.

9. _____ High sounds are made by fast vibrations.

10. _____ Sounds travel in waves.

Write the correct word from the list on each line.

directions	sound	animals
ears	people	larynx
weak	hearing aid	things
water	slow	strong
fast	eardrum	

11. Sounds are made by a._____, b._____, or c._____.

12. When something vibrates, it makes a _____.

13. Sounds go out in all directions like the rings made when you throw a stone in _____.

14. Sounds move in all _____.

15. Low sounds are made by _____ vibrations.

16. Sound goes into the ear and makes the _____ vibrate first.

17. People who cannot hear well may find help with a _____ _____.

18. When you make sounds, your _____ vibrates.

19. Loud sounds are made by _____ sound waves.

20. You hear with your _____.

Draw a line from the beginning to the correct ending of each sentence.

21. Soft sounds are made when   sound is made

22. When something vibrates,   vibrate when sound hits them.

23. The bones in your middle ear   something vibrates weakly.

```
┌─────────┐
│ 20  /   │
│    /    │
│   / 25  │
└─────────┘
```

Date _____

Score _____

Name _____

Draw a line to match

1. revolve   tilt

2. orbit   **one rotation of the earth**

3. rotate   spin

4. axis   go around something

5. day   path

Circle the correct word for each sentence. Write the word on the line.

6. The earth's axis _____ be seen.

    can   cannot

7. Fall is _____ than summer.

    warmer   colder

8. When one part of the earth is light, the other part is _____.

    light   dark

9. Spring comes _____ winter.

    before   after

10. The earth's rotation makes the _____ different on the east coast than the west coast.

    time   season

11. The earth takes one _____ to revolve around the sun.

    year   day

12. The year with the extra day is _____ year.

    jump   leap

13. When the earth's axis is toward the sun, we have _____.

               winter   summer

14. The earth moves in a path called its _____.

               axis   orbit

15. The earth _____ around on its axis.

               revolves   rotates

Write the correct words in the blanks.

16. The four seasons are a._____,

     b._____, c._____,

     and d._____.

17. The four time zones in the United States are a._____,

     b._____, c._____,

     and d._____.

Write the correct word from the list on the line.

time	seasons	hours
night	calendar	South Pole
North Pole	zones	rotation

18. You tell the day and month on a _____.

19. The United States has four time _____.

20. The north end of the axis is called the _____.

21. A day has twenty-four _____.

22. When the earth revolves, the _____ change.

23. When the earth rotates, the _____ changes.

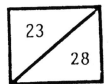

Date _____

Score _____

Name _____

Draw a line to match word with the correct phrase.

1. limestone   large, rounded stone

2. slates   created all matter

3. God   formed from limestone

4. core   used to write on

5. Mount Rushmore   carved figures on a mountain

6. geologist   sedimentary rock

7. boulder   wearing away of rocks and soil

8. pressure   center of the earth

9. erosion   helps in forming rocks

10. marble   studies surface of the earth

Answer *yes* or *no*.

11. _____ The name for the center of the earth is magma.

12. _____ Granite and basalt are igneous rocks.

13. _____ Nearly all minerals grow in a liquid.

14. _____ A geologist is a person who studies plants.

15. _____ Rocks are used for building.

Write the correct answer from the list on each line.

magma   water   igneous   plants   geologist

gems   volcano   pressure   minerals   hounds

16. Lava is melted rock that flows from a _____.

17. Granite is a very hard rock formed from _____.

18. Most rocks are made up of small crystals called _____

_____.

19. The high priest wore a breastplate of twelve _____

_____.

20. Granite is an example of _____ rock.

21. Rocks can be broken by pressure and _____.

22. People who gather rocks for fun are called rock _____

_____.

23. Rocks are changed by wind and _____.

24. Rocks are formed by heat and _____.

25. A person who studies about rocks is called a _____

_____.

20 / 25

Date _____

Score _____

Name _____

Write the correct word from the list on each line.

molecules	absorbed	resources
fuel	expand	friction
pollution	fire	conduction

1. Heat energy is the energy of vibrating _____.

2. Gases _____ when they are heated.

3. The heat energy that needs fuel and oxygen is

   _____.

4. Fuels are natural _____.

5. Careless use of heat energy can cause _____.

6. Heat energy caused by rubbing is called

   _____.

7. Light energy is changed to heat energy when it

   is _____.

8. Heat energy moves through solids by _____.

9. Food is the _____ that helps produce
   heat energy in the body.

Use this list to answer the next three questions.

friction	the body	rise
fire	expand	conduction
radiation	convection	the sun
electricity		

10. Put an *X* beside two words that tell what happens to gas when it is heated.

11. Underline five causes of heat energy.

12. Circle three ways heat energy moves.

Write the correct letter and word on each line.

13. The study of living things and where they live is

    called _____.

    a. environment   b. ecology

14. Air pollution is a _____ of heat energy.

    a. benefit   b. problem

15. Pollution is _____.

    a. man-made   b. God-given

16. To get smaller means to _____.

    a. expand   b. contract

17. Cooked food and warm houses are _____
    of heat energy.

    a. problems   b. benefits

18. Everything around you is part of your _____.

    a. environment   b. ecology

19. Natural resources are _____.

    a. man-made   b. God-given

20. To take up more space is to _____.

    a. expand   b. contract

$$\boxed{\begin{array}{c} 22 \\ \diagup \\ 27 \end{array}}$$

Date _____

Score _____

Name _____

Draw lines to match.

1. rotation

2. larva

3. conscience

4. eardrum

5. invertebrate

6. igneous rock

7. proper food, rest, and exercise

8. vertebrate

9. heat energy

10. earth revolving

11. Leap Year

12. solid

without a backbone

only humans have

important for healthy bodies

with a backbone

picks up sound vibrations

causes night and day

vibrating molecules

366 days in the year

worm stage of insects

formed by heat and pressure

fixed size and shape

going around the sun

Write the correct words from the list on the lines.

exercise   water   soil   rest

air   solids   liquids   temperature

sunlight   food   minerals   gases

13. Name the three forms of matter.

a. _____

b. _____

c. _____

14. List four things that the body needs to live and grow.

a. _____

b. _____

c. _____

d. _____

15. List five things plants need to grow.

a. _____

b. _____

c. _____

d. _____

e. _____

Complete this activity.

16. What is heat energy.

_____

_____

_____

Write *true* or *false*.

17. _____ Sounds are vibrations.

18. _____ Roots make the plant food.

19. _____ Stems grow below the ground.

20. _____ Wind, water, and cold can change rocks.

21. _____ All matter has weight and takes up space.

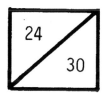

Date _____

Score _____

Notes

1. pumps blood
2. man has
3. in mouth
4. digestive system
5. openings in nose
6. breathe out
7. 4
8. 3
9. 2
10. 1
11. 5
12. Joseph Priestley
13. diaphragm
14. slower
15. oxygen
16. hairs
17. Examples; either order:
    a. Animals and people can breathe.
    b. Animals and people digest food.
18. Examples; any order:
    a. People have a spirit.
    b. People have a conscience.
    c. People can choose to think good thoughts (be creative, talk).
19. pulse
20. body
21. exercise
22. inhale
23. spirit
24. conscience

1. take in water and minerals
2. gas given off by plants
3. used to make things look larger
4. where some plants store food
5. take water and minerals to the leaves
6. chlorophyll
7. seed
8. carbon dioxide
9. food
10. roots
11. temperature
12. sprout
13. lettuce
14. apple
15. beans
16. celery
17. carrots
18. b
19. a
20. c
21. Any order:
    a. water
    b. minerals
    c. sunlight
    d. carbon dioxide
    e. right temperature

1. 212° F
2. 0° C
3. 32° F
4. thermometer
5. faster
6. vertebrates
7. potato bugs
8. expands
9. metamorphosis
10. mammals
11. Any order:
    a. temperature
    b. water
    c. light
    d. soil
12. Any order:
    a. fish
    b. amphibians
    c. reptiles
    d. birds
    e. mammals
13. Any order:
    a. egg
    b. larva
    c. pupa
    d. adult

1. water
2. God
3. energy
4. grow
5. junk
6. shoulder
7. grow
8. clothes
9. clean
10. sick
11. Any order:
    a. milk and dairy products
    b. meat and eggs
    c. bread and cereal
    d. fruits
    e. vegetables
    f. fats, oils, sweets
12. true
13. false
14. true
15. false
16. true
17. true

1. physical change
2. made all matter
3. changes shape and size
4. chemical change
5. tiny bits of matter
6. cannot be seen
7. fixed shape and size
8. space used by matter
9. how much matter is in something
10. a person that studies chemistry
11. chemical change
12. physical change
13. solid
14. yes
15. no
16. yes
17. yes
18. yes
19. yes
20. no

1. no
2. no
3. no
4. yes
5. no
6. yes
7. yes
8. no
9. yes
10. yes
11. Any order:
    a. people
    b. animals
    c. things
12. sound
13. water
14. directions
15. slow
16. eardrum
17. hearing aid
18. larynx
19. strong
20. ears
21. something vibrates weakly.
22. sound is made.
23. vibrate when sound hits them.

1. go around something
2. path
3. spin
4. tilt
5. one rotation of the earth
6. cannot
7. colder
8. dark
9. after
10. time
11. year
12. leap
13. summer
14. orbit
15. rotates
16. Any order:
    a. summer
    b. fall
    c. winter
    d. spring
17. Any order:
    a. Eastern
    b. Central
    c. Mountain
    d. Pacific
18. calender
19. zones
20. North Pole
21. hours
22. seasons
23. time

1. sedimentary rock
2. used to write on
3. created all matter
4. center of the earth
5. carved figures on a mountain
6. studies surface of the earth
7. large, rounded stone
8. helps in forming rocks
9. wearing away of rocks and soil
10. formed from limestone
11. no
12. yes
13. yes
14. no
15. yes
16. volcano
17. magma
18. minerals
19. gems
20. igneous
21. plants
22. hounds
23. water
24. pressure
25. geologist

1. molecules
2. expand
3. fire
4. resources
5. pollution
6. friction
7. absorbed
8. conduction
9. fuel
10. expand X
    rise X
11. <u>friction</u>
    <u>fire</u>
    <u>the body</u>
    <u>the sun</u>
    <u>electricity</u>
12. radiation
    convection
    conduction
13. b. ecology
14. b. problem
15. a. man-made
16. b. contract
17. b. benefits
18. a. environment
19. b. God-given
20. a. expand

1. causes night and day
2. worm stage of insects
3. only humans have
4. picks up sound vibrations
5. without a backbone
6. formed by heat and pressure
7. important for healthy bodies
8. with a backbone
9. vibrating molecules
10. going around the sun
11. 366 days in the year
12. fixed size and shape
13. Any order:
    a. solids
    b. liquids
    c. gases
14. Any order:
    a. air
    b. food
    c. water
    d. exercise
    e. rest
15. Any order:
    a. temperature
    b. water
    c. minerals
    d. soil
    e. sunlight
16. the energy of vibrating molecules
17. true
18. false
19. false
20. true
21. true

ANSWER KEYS

Section One

1.1 Example: The person who has been running used his supply of oxygen. He used his breath that had the oxygen.

1.2 Example: The body cannot get the oxygen it needs when air is not inhaled.

1.3 air

1.4 lungs

1.5 hairs

1.6 blood

1.7 warm

1.8 windpipe

1.9 a. nostrils
b. windpipe
c. two

1.10 The sponge became smaller.

1.11 The sponge became its size again.

1.12 yes

1.13 yes

1.14 My lungs were bigger because they had air in them.

1.15 yes

1.16 yes

1.17 no

1.18 yes

1.19 yes

1.20 yes

1.21 To inhale means to bring air into the lungs.

1.22 To exhale means to let the air out of the lungs,

1.23 Oxygen is taken from the air that is inhaled.

1.24 Carbon dioxide is exhaled.

1.25 1. windpipe
2. create
3. lungs
4. oxygen
5. nitrogen
6. carbon dioxide
7. nostrils

Section Two

2.1 Food is chewed and made soft.

2.2 Saliva makes food soft.

2.3 A little door closes off the windpipe.

2.4 food is broken up

2.5 very strong

2.6 more digestion

2.7 changing food for body use

2.8 Some things pass through paper while other things do not pass through paper.

2.9 The blood takes food to all parts of the body.

2.10 The food the body cannot use is pushed out as waste.

2.11 The food goes into the blood to all parts of the body or goes out as waste.

2.12 a. mouth
b. teeth
c. tongue
d. long tube
e. stomach
f. small intestine
g. blood
h. large intestine

Section Three

3.1 Any order:
   a. air
   b. food
   c. water
   d. exercise
   e. rest

3.2 faster

3.3 moves

3.4 faster

3.5 oxygen

3.6 Any order: oxygen, food, water

3.7 exercise

3.8 teacher check

3.9 consonant

3.10 vowel

3.11 digging   skipping
     flopping   hugging
     grabbing   swimming

3.12 Examples:
     I am running fast.
     Joe is digging a hole.
     He was flopping into the chair at
     one o'clock.
     Jane was skipping rope.
     Ann was hugging her mother.

Bob was swimming in the pool.
Tom was grabbing the rail.

3.13 Examples:
   a. playing ball   d. biking
   b. jumping rope   e. hiking
   c. playing tag   f. running

3.14 teacher check

3.15 brain

3.16 yes

3.17 yes

3.18 no

3.19 yes

3.20 yes

3.21 yes

3.22 yes

3.23 protects brain

3.24 carries messages

3.25 supports whole body

3.26 where two bones meet

3.27 206

3.28 needed for a strong body

3.29 Answers may vary.

Section Four

4.1   third

4.2   thirsty

4.3   under

4.4   church

4.5   purse

4.6   spirit

4.7   instinct

4.8   spirit

4.9   mind

4.10   creative

4.11   changing

4.12   measure

4.13   Any order

Same	Different
water	conscience
breathing system	mind
food	spirit
bones	
muscles	
air	
life	
digestive systems	

Section One

1.1	teacher check	1.8	food	
1.2	roots	1.9	carbon dioxide	

1.3   Either order: a. stems b. leaves

1.4   Either order: a. water b. minerals

1.5   Example:
I saw the colored water in the stem of the celery stem. Some of the colored water also gets into the leaves.

1.6   stem

1.7   leaves

1.10   a. water   b. oxygen

1.11   God

1.12   teacher check

1.13   a. taking water and minerals to the leaves

b. taking in carbon dioxide or making food

Section Two

2.1   The plant that was not covered.

2.2   It did not get the sunlight to make chlorophyll.

2.3   It would look pale and may be dead.

2.4   a. water   b. minerals

2.5   carbon dioxide

2.6   It will freeze and die.

2.7   Any order:
a. water   b. minerals

c. sunlight   d. carbon dioxide
e. right temperature

2.8   a. **root**   b. **look**

tooth	cook
room	foot
soon	hook
tool	wood
school	good
boot	stood

Section Three

3.1 – 3.3   teacher check

3.4   a. a tiny plant
b. food for the tiny plant

3.5   water

3.6   no

3.7   a. cattails   b. strawberry
c. white potato

3.8   stem

3.9   cutting

3.10   Some parts grow faster than others

3.11   the stem

Section One

1.1	temperature		1.11	soil, water
1.2	Either order:		1.12	soil, water
	100° Celsius, 212° Fahrenheit		1.13	land, ocean
1.3	water, ice		1.14	teacher check
1.4	have heavy fur or thick fat		1.15	temperature, water
1.5	teacher check		1.16	teacher check
1.6	teacher check		1.17	teacher check
1.7	Answers may vary.		1.18	temperature, water, light
1.8	The ice cubes are higher than the water was.		1.19	teacher check
1.9	teacher check		1.20	Any order: temperature, light, water, soil
1.10	teacher check			

Section Two

2.1 Examples:

1. dog	6. snake
2. cat	7. zebra
3. horse	8. elephant
4. pig	9. monkey
5. cow	10. fish

2.2 Examples:
1. size
2. shape
3. color
4. where they live

Examples:

2.3 8

2.4 0

2.5 1

2.6 1

2.7 no

2.8 10

2.9 Any order:

a. fish	d. birds
b. amphibians	e. mammals
c. reptiles	

2.10 invertebrates

2.11 95

2.12 vertebrates or invertebrates

2.13 teacher check

2.14 teacher check

2.15

Warm-blooded	**Cold-blooded**
mammals	fish
birds	reptiles
	amphibians

2.16

Gill-Breathers	**Lung-Breathers**
fish	reptiles
amphibians	amphibians
	birds
	mammals

2.17

From Eggs	**Born Alive**
fish	mammals
amphibians	
reptiles	
birds	

2.18

1. spider	6. worm
2. ant	7. fly
3. grasshopper	8. bee

4. roach   9. wasp

5. lady bug   10. snail

2.19   7

2.20   1

2.21   2

2.22   Examples: clams, lobsters, shrimp

2.23

Plant Eaters	**Plant & Meat Eaters**
zebra	
rabbit	man
cow	bear
horse	woodpecker

Meat Eaters
lion
hawk
owl
dog
mountain lion

2.24   teacher check

Section Three

3.1   egg, pupa, larva, adult

3.2   shed or molted

3.3   1. egg   2. larva   3. pupa   4. adult

3.4   teacher check

3.5 – 3.6   teacher check

3.7   brown, owl, town, clown

3.8   round, thou, pouch, out

3.9   oy, oi

3.10

-ing	**-ed**
living	lived
breathing	breathed
measuring	measured

3.11   mammal, adult, pupa, larva

3.12   mammal, insect, amphibian, spider

Section One

1.1   a. John got out of bed.
      b. John prayed at breakfast.
      c. John ran to school.
      d. John said, "Good morning, Mrs. Farmer."
      e. Mrs. Farmer wrote on the board.

1.2   teacher check

1.3   Answers will vary.

1.4   milk

1.5   nutrients

1.6   two, three

1.7   Answers will vary.

1.8   1, #1

1.9   1, #1

1.10   3, #4

1.11   energy

1.12   teeth, bones

1.13   Answers will vary.

1.14   3, #5

1.15   3, #4

1.16   1, #1

1.17   Drawings will vary.

1.18

Level	Group #	Name
1	#1	bread, cereals
2	#2	vegetables
2	#3	fruits
3	#4	milk, yogurt, cheese
3	#5	meats, eggs
4	#6	fats, oils, sweets

1.19   meat and eggs

1.20   milk

1.21   milk

1.22   fruits

1.23   bread and cereal

1.24   meat and eggs

1.25   The food doesn't help the body.

1.26   no

1.27   Examples: carrots, cheese, celery, apples, peaches

1.28   an additive

1.29   an additive

1.30   an additive

Section Two

2.1 – 2.5   teacher check

2.6   none

2.7   nutrients

2.8 – 2.12   teacher check

2.13   corn
      more
      oranges
      for

2.14   pant
      jar
      car
      bar

2.15   teacher check

Section Three

3.1 milk, cheese, leafy vegetables

3.2 well

3.3 brush

3.4 grow

3.5 up

3.6 down

3.7 twice

3.8 floss

3.9 tool

3.10 Examples: (any four; any order)

I am going to take good care of my eyes. I will not keep reading when the room is getting dark. I will wear sunglasses when I am in the bright sunshine. I will eat foods like carrots. They are rich in Vitamin A and that is good for the eyes. I am going to be careful when using sharp pointed tools. I will not point sharp tools toward my eyes or those of my friends.

3.11 no

3.12 no

3.13 yes

3.14 yes

3.15 no

3.16 yes

3.17 let/tuce

mid/dle

scis/sors

rab/bit

hap/py

ap/ple

3.18 teacher check

Section One

1.1 Drawings will vary.

1.2 what things are made of

1.3 chemistry

1.4 Robert Boyle

1.5 yes

1.6 He had the Bible printed in Irish and Gaelic

1.7 a chemist

1.8 Either order:

    a. He tries to find out what is in matter.

    b. He puts different kinds of matter together.

1.9 clear

1.10 cloudy, dirty, not clear

1.11 yes

1.12 chemist

1.13 dirty, cloudy water, and sand

1.14 clear water, clean water

1.15 sand, wet sand

1.16 yes

1.17 sand and water

1.18 chemist

1.19 Any order:

rock	ruler	tomato
hard	long	soft
rough	hard	smooth
heavy	light	round

1.20 true

1.21 true

1.22 true

1.23 false

1.24 teacher check

1.25 desk, water, air, glass, pencil

1.26 teacher check

1.27 teacher check

1.28 The objects with the biggest volume take up the most space.

1.29 falls to the ground

1.30 falls down

1.31 jump up and come down

1.32 brick

1.33 The brick has more mass than the book. The force of gravity pulls more on the brick.

1.34 No, the hanger tipped down.

1.35 The end where the blown-up balloon hung.

1.36 The blown-up balloon has more mass.

1.37 yes

1.38 yes

1.39 yes

1.40 teacher check

1.41 yes

1.42 Water has mass and takes up space.

1.43 yes

1.44 yes

1.45 Mass is the amount of matter in an object.

Section Two

2.1--2.6 Answers will vary.

2.7   yes

2.8   yes

2.9   no

2.10   Example: wood

2.11   Example: milk

2.12   Example: oxygen

2.13   fixed size and fixed shape

2.14   fixed size and changing shape

2.15   changing shape and changing size

2.16   no

2.17   yes

2.18   not water (B)

2.19   The hot water melted the ice cube.

2.20   (A)

2.21   one in freezer (E)

2.22   The ice cube in the freezer is too cold to melt. Not enough heat was available.

2.23   no

2.24   melt

2.25   liquids

2.26   gases

2.27   solids

2.28   hear

2.29   cheer

2.30   chair

2.31   fair

2.32   pear

2.33   fear

2.34   hair

2.35   deer

2.36   yes

2.37   yes

2.38   yes

2.39   no

2.40   yes

2.41   Example: Break a rock into pieces.

2.42   When the vinegar touched the baking soda, bubbles formed. The baking soda was eaten away. (chemical change)

2.43   solid

2.44   liquid

2.45   gas

2.46   chemical change

2.47   Examples:
a. digesting food   d. burning
b. baking a cake   e. rusting
c. food spoiling   f. plant growing

2.48   "Jesus Christ the same yesterday, and today, and forever."

2.49   gas

2.50   physical change

2.51   chemical change

2.52   temperature

2.53   solid

2.54   a. atoms   b. molecules

2.55   liquid

Section One

1.1 Examples:
moving feet,
teacher talking,
heater, whispering

1.2 Examples:
cars, horns, birds singing

1.3 Examples:
Kitchen — pots and pans,
voices, radio
Living
room — voices, piano,
television
Bedroom– radio, bird
singing, voices,
dog

1.4 Examples:
<u>People</u>
teacher
children
voices
moving feet

<u>Animals</u>
birds
dog
<u>Things</u>

radio
pots and pans
piano
heater

1.5 people

1.6 things

1.7 animals

1.8 a. people
b. animals
c. things

1.9 yes

1.10 no

1.11 yes

1.12 top of the drum moved

1.13 no

1.14 yes

1.15 moving

1.16 fast

1.17 vibrate

1.18 vibrations

1.19 han/dle   a/ble   cra/dle
grid/dle   sta/ble   fable
la/dle   am/ple   cou/ple

Section Two

2.1 rings in water

2.2 outward in circles

2.3 yes

2.4 yes

2.5 yes

2.6 <u>Down to him</u>
helicopter

bird
<u>Up to him</u>
dog
<u>Across to him</u>
truck
jumping girl

2.7 yes

2.8 yes

2.9   yes

2.10   yes

2.11   a. yes
       b. Sound waves move out
          in all directions.

2.12   joyful

2.13   God speaking to people

2.14   face

2.15   yes

2.16   yes

2.17   yes

2.18   no

2.19   a. (hear)

2.20   b. (had your ear to the glass)

2.21   a. (do)

2.22   b. (do not)

2.23   larger rubber bands

2.24   smaller rubber bands

2.25   smaller

2.26   faster

2.27   slower

2.28   larger

2.29   smaller pan

2.30   smaller pan

2.31   larger pan

2.32   faster vibrations

2.33   slower vibrations

2.34   The cereal hopped around a little
       bit.

2.35   The cereal jumped around alot.

2.36   When I hit it hard.

2.37   fast

2.38   slow

2.39   strong

2.40   weak

2.41   Any order:
       wheelbarrow
       mailman
       catnip
       wristwatch
       Sunday
       doorknob

2.42   Examples:
       into
       friendship
       afternoon

Section Three

3.1 I could hear better with my ears uncovered.

3.2 tube of paper

3.3 the outside part

3.4 outer ear

tube

eardrum

three bones in middle ear

inner ear

3.5 teacher check

3.6 The cereal hopped around.

3.7 The sound waves made it vibrate.

3.8 outer ear → eardrum →

the bones of the middle ear →

inner ear → nerves → brain

3.9 (larynx), (larynx), (larynx),

(larynx)

3.10 [air]

3.11 When the larynx vibrates, you make sounds.

3.12 hearing aid

3.13 stronger

3.14 nerves

3.15   a. people
  b. sound
  c. walk
  d. light

3.16   a. han dle
  b. sta ble
  c. a ble
  d. ta ble

Section One

1.1   dark

1.2   light

1.3   Examples:
United States
Mexico
Hawaii
Canada
Greenland
North Pole

1.4   Examples:
China
Russia
Europe
Africa
India
France

1.5   Examples:
China
Russia
Europe
Africa
India
France

1.6   yes

1.7   day

1.8   night

1.9   day

1.10   night

1.11   teacher check

1.12   untold   not told

1.13   unfair   not fair

1.14   unusual   not usual

1.15   unwelcome   not welcome

1.16   unfriendly   not friendly

1.17   sleep

1.18   go to school, play, and eat

1.19   sleeping

1.20   no

1.21   yes

1.22   no

1.23   My body keeps the light from shining on my face.

1.24   The other side of me would be dark.

1.25   The earth is in the way.

1.26   yes

1.27   North Pole / South Pole (globe diagram)

1.28   rotates — twenty - four hours

1.29   axis — a model of the earth

1.30   North Pole — a handle through the earth

1.31   globe — bottom of the globe

1.32   South Pole — turns around

1.33   a day — top of the globe

1.34   teacher check

1.35   r o t a t e s

1.36   o r b i t

1.37   a x i s

1.38   r e v o l v e s

1.39   t r a v e l s

1.40   teacher check

Section Two

2.1

2.2

2.3   6

2.4

2.5   6

2.6   12

2.7

2.8   6

2.9   18

2.10

2.11   2

2.12   20

2.13

2.14   4

2.15   24

2.16   12

2.17

	Month	Days
a.	January	31
b.	February	28
c.	March	31
d.	April	30
e.	May	31
f.	June	30
g.	July	31
h.	August	31
i.	September	30
j.	October	31
k.	November	30
l.	December	31

2.18   365

2.19   teacher check

2.20   teacher check

2.21   no

2.22   no

2.23   yes

2.24   The time zone was different.

2.25   24

2.26   4

2.27   8 o'clock

2.28   7 o'clock

2.29   4

2.30   a. Eastern
        b. Central
        c. Mountain
        d. Pacific

2.31   sleeping

Section Three

3.1
   a. together
   b. Lord
   c. time
   d. kingdom
   e. times
   f. seasons
   g. power

3.2   Times will vary.

3.3   Pictures will vary.

3.4   Times will vary.

3.5   Pictures will vary.

3.6   no, only north

3.7   North Star

3.8   north axis of the earth

3.9   John went to baseball practice

3.10   They went into the store.

3.11   no

3.12   yes

3.13   yes

3.14   no

3.15   yes

3.16   pictures will vary

3.17   Examples:
   a. Christmas
   b. New Year
   c. Valentine's Day
   d. skiing
   e. sledding
   f. snowmen

3.18   hotter

3.19   to

3.20   more

3.21   more

3.22   before

3.23   Pictures will vary.

3.24   Examples:
   a. swimming
   b. boating
   c. vacation
   d. company
   e. fishing
   f. 4th of July

3.25   light

3.26   sun

3.27   heat

3.28   teacher check

3.29   Pictures will vary.

3.30   Examples:
   a. leaves on trees
   b. birds come back
   c. blossoms on trees
   d. rain
   e. Easter
   f. flowers

3.31   after

3.32   warmer

3.33   colder

3.34   before

3.35   teacher check

3.36   Pictures will vary.

3.37   Examples:
   a. school starts
   b. Thanksgiving Day
   c. pumpkins
   d. leaves are colored
   e. leaves fall
   f. football

3.38   summer

3.39   colder

3.40   warmer

3.41   before

3.42   replay — play again

3.43   review — view again

3.44   reweigh — weigh again

3.45   rewind — wind again

3.46   remake — make again

3.47   disable — not able

3.48   disappear — to not appear

3.49   disapprove — to not approve

3.50   discomfort — not in comfort

3.51   dishonor — to not honor

3.52   dislike — to not like

Section One

1.1 geologists

1.2 igneous

1.3 hounds

1.4 hard

1.5 a. hike
   b. rocks

1.6 God

1.7 hard or igneous

1.8 rock

1.9 magma

1.10 crust

1.11 a. mantle
   b. core
   c. crust

1.12 a. center
   b. liquid rock

1.13 a. middle
   b. rock

1.14 a. top layer
   b. soil
   c. rocks

1.15 a. rock
   b. plants
   c. animals

1.16 liquid

1.17 volcano

1.18 magma

1.19 basalt

1.20 blow

1.21 lava

1.22 Either order:
   a. Heat
   b. Pressure

1.23 true

1.24 true

1.25 false

1.26 false

1.27 true

1.28 false

1.29 true

1.30 true

1.31 true

1.32 teacher check

1.33 The grains of rock bubbled a little.

1.34 There was a fizzing noise when I dropped the vinegar on the rock.

1.35 I used vinegar to make the grains bubble.

1.36 teacher check

1.37 a. no
   b. yes
   c. yes
   d. no
   e. yes
   f. yes
   g. no

1.38 six

1.39 cubes

1.40 water

1.41 liquid

1.42 liquid

1.43 changed

1.44 granite

1.45 metamorphic

1.46 hard

1.47 heat and pressure

Section Two

2.1   water

2.2   smaller

2.3   water

2.4   water

2.5   earth

2.6   Either order:
    a.   wind
    b.   water

2.7   sandstone

2.8   Either order:

    a. water
    b.   wind

2.9   slowly

2.10   sand

2.11   slowly

2.12   change

2.13   break into chunks

2.14   freezes

2.15   four

2.16   a.   slower, slowest, slowly
    b.   cooler, coolest
    c.   softer, softest, softly
    d.   warmer, warmest, warmly
    e.   larger, largest, largely

2.17   a.   funniest, funnier
    b.   crazier, crazily
        craziest, craziness
    c.   greediest, greedier
        greedily, greediness
    d.   happiest, happily
        happier, happiness
    e.   fanciest, fancier
        fancily

2.18   teacher check

2.19   Any order:
    a. world
    b. wind
    c. force
    d. cracks
    e. roots
    Any order:
    f. visit
    g. uncle
    h. water
    i. boulder
    j. college
    k. question
    l. crumble

2.20   true

2.21   false

2.22   true

2.23   false

2.24   true

2.25   true

2.26   true

2.27   flowers

2.28   living

2.29   uncle

2.30   little

2.31   college

2.32   boulders

Section Three

3.1   teacher check

3.2   stones

3.3   land

3.4   slate

3.5   minerals

3.6   true

3.7   true

3.8   false

3.9   true

3.10   true

3.11   false

3.12   false

3.13   Example:

This house is made of limestone.

3.14   Any order:
- a. George Washington
  Thomas Jefferson
  Theodore Roosevelt
  Abraham Lincoln
- b. South Dakota
- c. Gutzon Borglum
- d. drills and dynamite
- e. as high as a 5 story building (about 60 feet)
- f. yes

3.15   yes

3.16   no

3.17   yes

3.18   yes

3.19   yes

3.20   no

3.21   rock hounds

3.22   egg

3.23   labeled

3.24   teacher check

3.25   Any order:
- a. sardius (ruby)
- b. topaz
- c. carbuncle (garnet)
- d. emerald
- e. sapphire
- f. diamond
- g. ligure (turquoise)
- h. agate
- i. beryl
- j. onyx (carnelian)
- k. jasper
- l. amethyst

3.26   a. tribes

3.27   b. breastplate

3.28   b. gold

3.29   b. polished

3.30   c. diamond

3.31   c. minerals

3.32   b. gems

3.33   a. diamonds

Section One

1.1 warm

1.2 warm

1.3 cold

1.4 hot

1.5 friction

1.6 Examples:
Striking a match on a stone or a matchbox will start a fire. Indians used to start fires by rubbing two sticks together.

1.7 yes

1.8 no

1.9 yes

1.10 yes

1.11 Solid Fuel
wood
coal

Liquid Fuel
oil
gasoline

Gas Fuel
natural gas
butane

1.12 Examples
candle under pint jar
   The flame went first.
candle under quart jar
   The flame burned a little longer.
uncovered candle
   The flame kept burning.

1.13 Example:
The oxygen in the smaller jar was burned up first.
The oxygen in the quart jar lasted a little longer.
The uncovered candle could keep getting oxygen from the air.

1.14 Example: teacher check:
Ashes and cinders can be used to keep cars from sliding on slippery streets.

1.15 teacher check

1.16 d dynamic — a. can be turned into electricity

1.17 c static — b. used to sent current

1.18 a moving water — c. electricity found in nature

1.19 b wires — d. electricity produced by man

1.20 a. heat or warm air
b. Example:
   the heat comes from inside my body

1.21 teacher check

1.22 teacher check

1.23 Example:
yes / no

1.24 Example:
to get warm

1.25 Example:
Light energy goes right through the transparent glass. It is absorbed by the cat. The cat feels warm.

1.26 yes
Example:
The glass might have felt a little warmer.

1.27 yes
Example:
The sunlight shining through the glass burned a hole in the paper.

1.28 Example:
The magnifying glass is transparent and the paper is not. Light energy was absorbed by the paper to make heat energy.

1.29 Any order:

a. friction
b. electricity
c. fire
d. our bodies
e. the sun

Section Two

2.1 teacher check

2.2 hot

2.3 vibrate

2.4 heat energy

2.5 molecules

2.6 air

2.7 gas

2.8 expands

2.9 gets bigger

2.10 move apart

2.11 <u>GAS</u>
fog
steam
cloud

<u>LIQUID</u>
raindrop
dew
ocean

<u>SOLID</u>
snow
ice cube
hail

2.12 Example:
The ice cubes would melt.

2.13 heat energy

2.14 1 cup

2.15 Example:
The water was moving very fast. Steam was rising.

2.16 an amount less than one cup

2.17 Example:
The water turned into steam and got mixed up with the air.

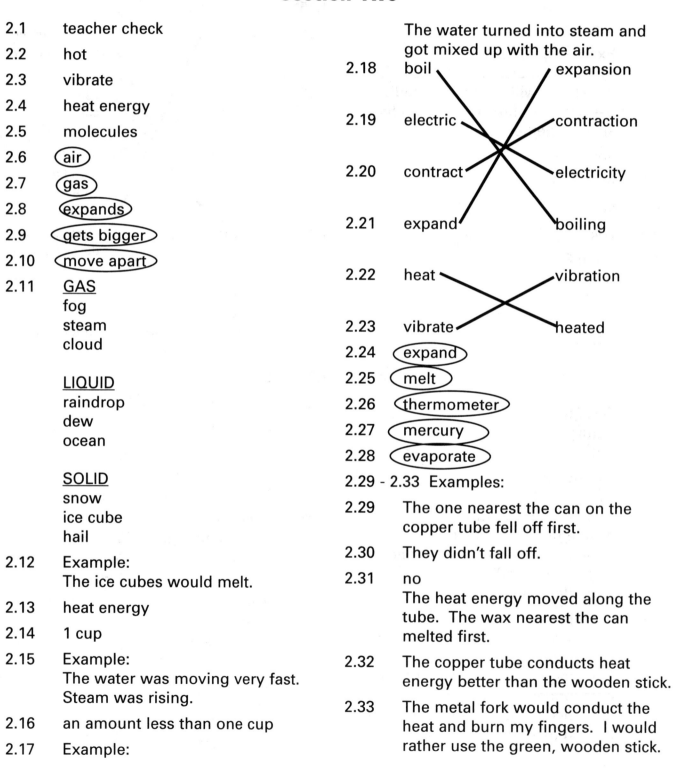

2.18 boil — boiling
2.19 electric — electricity
2.20 contract — expansion
2.21 expand — contraction

2.22 heat — heated
2.23 vibrate — vibration

2.24 expand

2.25 melt

2.26 thermometer

2.27 mercury

2.28 evaporate

2.29 - 2.33 Examples:

2.29 The one nearest the can on the copper tube fell off first.

2.30 They didn't fall off.

2.31 no
The heat energy moved along the tube. The wax nearest the can melted first.

2.32 The copper tube conducts heat energy better than the wooden stick.

2.33 The metal fork would conduct the heat and burn my fingers. I would rather use the green, wooden stick.

2.34   Examples:
        edge
        hedge
        wedge
        bridge
        ridge
        fudge
        badge

2.35   Any order:
        a. convection
        b. radiation
        c. conduction

Section Three

3.1   the sun

3.2   friction

3.3   electricity

3.4   your body

3.5   fire

3.6   CAUSES
        Any order:
        friction
        fire
        electricity
        your body
        the sun.

        BENEFITS
        Examples:
        It can warm my hands.
        It keeps my house warm.
        It cooks my food.
        It gives me energy to work and play.
        It keeps the earth warm.

3.7   Examples:
        Be sure campfires are out.
        Never throw a lighted match on the ground.
        Report any smoke or fire you might see.

3.8   Any order:
        a. wood
        b. oil
        c. gasoline
        d. coal
        e. natural gas

3.9   teacher check

3.10   Example:
        A description of sunburn with its related problems of itching, peeling, and hurting would be appropriate.

3.11   Any order:
        a. fish
        b. turtles
        c. clams
        d. sea weed
        e. reeds
        f. lily pads

3.12   Example:
        We need clean water to drink, to cook with, and to wash with every day. Polluted water is very dangerous and may make us sick.

3.13   teacher check

Section One

1.1 Any order:
 a. fruit
  Example: orange
 b. bread and cereal
  Example: corn flakes
 c. milk
  Example: milk and cheese
 d. meat and eggs
  Example: hamburger
 e. vegetables
  Example: broccoli
 f. fats, oils, and sweets
  Example: potato chips

1.2 Any order:
 a. air
 b. food
 c. exercise
 d. rest

1.3 inhale

1.4 air

1.5 windpipe

1.6 lungs

1.7 blood

1.8 carbon dioxide

1.9 instinct

1.10 creative

1.11 create

1.12 conscience

1.13 Either order:
 a. conscience
 b. spirit

1.14 Any order:
 a. true
 b. honest
 c. just
 d. pure
 e. lovely
 f. of good report

1.15
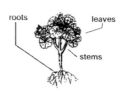

1.16 carbon dioxide

1.17 soil

1.18 die

1.19 seed

1.20 oxygen

1.21 Any order:
 a. water
 b. correct temperature
 c. mineral
 d. sunlight
 e. carbon dioxide

1.22

T	X	V	L	B	O	D
E	V	W	A	T	E	R
M	W	M	O	V	X	Y
P	M	N	M	S	W	T
E	N	W	O	U	M	N
R	P	I	B	N	S	U
A	L	T	A	L	I	G
T	Q	E	X	I	L	O
U	R	X	Y	G	Z	X
R	V	W	Z	H	V	Y
E	W	M	B	T	O	B

1.23   <u>Down</u>
    2. adult
    3. larva
    4. insect
    6. egg

<u>Across</u>
  1. pupa
  5. invertebrate
  7. starfish

Section Two

2.1   Examples; any order:
    a. red
    b. soft
    c. smooth
       or light

    d. M
    e. S
    f. I
    g. I
    h. S

2.2   solid

2.3   liquid

2.4   solid

2.5   gas

2.6   liquid

2.7   gas

2.8   teacher check

2.9   twenty-four

2.10 day

2.11 dark

2.12 9:00 p.m.

2.13 four

2.14 365 days

2.15 twenty-nine

2.16 one

2.17 salt

2.18 diamond

2.19 coal

2.20 a. I
      b. M
      c. S

2.21 teacher check

2.22

```
C O (H E A T) T I V (W
F C D L E Z N A O  I
K Q P O N D O W B  N
(P O L N I M V M P  D)
X L E D B C U B A  C
Y D A S E B O L C  Y
F V R N E R T T K  S
G G (W A T E R) V (C  Z
H J I Q B S V U  O  T
A S K T V U S Y  L  L
I H O W A T C F  D) E
(E C I) W E H V G F D
```

Section Three

3.1   stronger

3.2   vibrations

3.3   eardrum

3.4   brain

3.5   waves

3.6   Example:
like a chime or church bell or dong

3.7   through the hanger, string, and finger

3.8   a.   room temperature
b.   room temperature

3.9   a.   varies
b.   varies

3.10   yes

3.11   because dark colors absorb heat energy and light colors reflect heat energy

3.12   Any order:
a.   friction
b.   fire
c.   electricity
d.   your body
e.   sun

Self Test 1

1.01	lungs	1.011	cleans the air in nose
1.02	oxygen	1.012	breathing in
1.03	carbon dioxide	1.013	breathing out
1.04	nitrogen	1.014	air becomes this before going into lungs
1.05	nostrils	1.015	number of lungs
1.06	windpipe	1.016	lungs are like this
1.07	Joseph Priestley	1.017	man who studies life
1.08	body	1.018	a. windpipe b. lungs c. nostrils
1.09	air		
1.010	nose		

Self Test 2

2.01	saliva	2.012	3
2.02	stomach	2.013	1
2.03	oxygen	2.014	4
2.04	nostrils	2.015	2
2.05	windpipe	2.016	2
2.06	waste	2.017	1
2.07	lungs	2.018	3
2.08	mouth	2.019	Joseph Priestly found out about oxygen.
2.09	small	2.020	Digestion is changing food to use in the body.
2.010	blood		
2.011	5		

Self Test 3

3.01	no	3.07	no
3.02	yes	3.08	yes
3.03	no	3.09	no
3.04	no	3.010	no
3.05	no	3.011	h. framework
3.06	yes	3.012	e. mouth

3.013   c. breathing muscle

3.014   g. pumping muscle

3.015   b. heart beat

3.016   a. inhale

3.017   f. exhale

3.018   d. takes food to body

3.019   Exercise helps take care of the body systems.

3.020   Food is chewed and softened.

3.021   The heart pumps blood through the body.

3.022   The spinal cord carries messages to the brain from the body.

3.023   The bones support and protect the body.

3.024   Digestion is changing food so the body can use it.

3.025   The pulse is the pumping of the blood you can feel in some spots of your body.

Self Test 4

4.01   4

4.02   3

4.03   1

4.04   5

4.05   2

4.06   no

4.07   yes

4.08   no

4.09   yes

4.010   yes

4.011   no

4.012   no

4.013   no

4.014   yes

4.015   no

Examples:

4.016   Animals and people can breathe.

4.017   Animals and people need air, water, and food.

4.018   Animals and people digest food.

4.019   People have consciences.

4.020   People have spirits.

Self Test 1

1.01   take in water and minerals

1.02   make and store food

1.03   material that is not plant or animal

1.04   plant's tiny pipes

1.05   thing used to make things look larger

1.06   take in carbon dioxide

1.07   gas taken from the air by plants

1.08   gas needed by animals and people to breathe

1.09   to cause to look larger

1.010   ground; earth; dirt

Any order:
1.011   a. roots   b. stems   c. leaves

1.012   a. water   b. minerals   c. carbon dioxide

1.013   a. microscope   b. magnifying glass

1.014   air

1.015   a. beets   b. carrots

1.016   oxygen

1.017   root hairs

1.018   God

1.019   Green plants give off oxygen and make food for people

Self Test 2

2.01   green coloring in plants

2.02   material in the soil needed by plants

2.03   thing by which temperature is measured

2.04   gas in the air needed by people

2.05   plant parts

2.06   b.   carbon dioxide

2.07   a.   soil

2.08   a.   freeze

2.09   a.   the beet

2.010   b.   down

Any order:
2.011   sunlight

2.012   carbon dioxide

2.013   water

2.014   minerals

2.015   right temperature

2.016   roots

2.017   stems

2.018   leaves

2.019   seeds

2.020   fruit

Self Test 3

3.01	tiny part of a plant that is just beginning to grow		3.010	oxygen
3.02	plant that sends out stems to make new plants		3.011	chlorophyll
3.03	a bulb we eat		3.012	white potato
3.04	a seed we eat		3.013	bulb
3.05	a thing used to make small things look larger		3.014	freeze
3.06	seed		3.015	green plants
3.07	roots		3.016	water, warmth
3.08	stem		3.017	roots, stems, seeds
3.09	grass or grass plants		3.018	a. leaves
			3.019	b. stem
			3.020	c. roots

Self Test 1

1.01	0^oC	1.05	soil
1.02	32^oF	1.06	Any order: temperature, water, light, soil
1.03	expands	1.07	temperature
1.04	soil		

Self Test 2

2.01 boiling point of water

2.02 backbones

2.03 six legs

2.04 butterfly

2.05 dog

2.06 Any order:
fish, amphibians, reptiles, birds, mammals

2.07 temperature, water, soil, light

2.08 a. antenna
b. head
c. thorax
d. abdomen

Self Test 3

3.01 reptile

3.02 fish

3.03 invertebrates

3.04 shed

3.05 amphibians

3.06 pupa

3..07 $32°F$

3.08 invertebrates

3.09 temperature

3.010 is not

3.011 amphibians

3.012 a. egg   b. larva
c. pupa   d. adult

3.013 Any order:
temperature, light, water, soil

3.014 Any order:
fish, amphibians, reptiles, birds, mammals

Self Test 1

1.01   A|FRUITS|NAD|VEGETABLES|Z

1.02   DXFJL|MILK|O|FATSANDOILS|

1.03 |BREADANDCEREAL|LIMSTW

1.04   MJNTH|MEATANDEGGS|INDV

1.05–1.10 In any order

1.05   fruits

1.06   milk, yogurt, cheese

1.07   bread and cereal

1.08   meat and eggs

1.09   vegetables

1.010   fats, oils, sweets

1.011   milk and cheese

1.012   meat and eggs

1.013   fats and oils

1.014   bread and cereal

1.015   meat and eggs

1.016   meat and eggs

1.017   fruits

1.018   bread and cereal

1.019   milk and cheese

1.020   fruits

1.021   vegetables

1.022   additives

1.023   meat and eggs

1.024   scurvy

1.025   bread and cereal

1.026   nutrients

1.027   energy

1.028   junk

Self Test 2

2.01   fruits and vegetables

2.02   meat and eggs

2.03   bread and cereal

2.04   milk

Any order:

2.05   fruits

2.06   meat and eggs

2.07   bread and cereal

2.08   milk

2.09   vegetables

2.010   fats, oils, sweets

Examples:

2.011   milk

2.012   hamburger

2.013   beans

2.014   apple

2.015   cheese

2.016   salad

2.017   milk

2.018   chicken

2.019   potato

2.020   bread

2.021   corn

2.022   ice cream

Self Test 3

3.01   twice

3.02   sun

3.03   carrots

3.04   down

3.05   day

3.06   hair

3.07   grow

3.08   clean

3.09   water

3.010   junk

Any order:

3.011   fruits

3.012   meat and eggs

3.013   bread and cereal

3.014   milk, yogurt, cheese

3.015   vegetables

3.016   fats, oils, sweets

Examples:

3.017   cheese   milk

3.018   sandwich   bread and cereal

3.019   apple   fruits

3.020   milk   milk

Self Test 1

1.01	tell about something	1.011	yes
1.02	worked on matter	1.012	yes
1.03	is a chemist	1.013	yes
1.04	is how much matter is in it	1.014	yes
1.05	pulls to center of the earth	1.015	no
1.06	no	1.016	no
1.07	yes	1.017	no
1.08	yes	1.018	yes
1.09	yes	1.019	no
1.010	yes	1.020	yes

Self Test 2

2.01	has fixed shape and size	2.011	chemical change
2.02	has size but shape changes	2.012	physical change
2.03	shape and size change	2.013	physical change
2.04	cannot be seen	2.014	chemical change
2.05	smallest bits of matter	2.015	chemical change
2.06	melting an ice cube	2.016	yes
2.07	burning wood	2.017	no
2.08	space used	2.018	yes
2.09	a study of matter	2.019	yes
2.010	what things are made of	2.020	yes

Self Test 1

1.01	moves fast		1.09	moved
1.02	vibrate		1.010	sound
1.03	people		1.011	yes
1.04	an animal		1.012	yes
1.05	moved		1.013	no
1.06	things		1.014	yes
1.07	vibrated		1.015	no
1.08	people			

Self Test 2

2.01	directions		2.016	yes
2.02	waves		2.017	no
2.03	things		2.018	yes
2.04	a. people		2.019	no
	b. animals		2.020	no
2.05	fast			
2.06	slow			
2.07	vibrations			
2.08	strong			
2.09	weak			
2.010	vibrating			
2.011	a. a louder			
2.012	b. lower			
2.013	b. vibrates faster			
2.014	a. strong			
2.015	c. weak			

Self Test 3

3.01 You hear with ————————— all directions.

3.02 A sound is made when ———— by fast vibrations.

3.03 Sounds come from ————————— something vibrates.

3.04 When you talk ———————————— your larynx vibrates.

3.05 Low sounds are made ————— your ears.

by slow vibrations.

3.06	(air)	3.021	yes
3.07	(hearing aid)	3.022	yes
3.08	(eardrum)	3.023	no
3.09	(brain)	3.024	no
3.010	(people, animals and things)	3.025	yes
3.011	(weak)		
3..012	(waves)		
3.013	(a sound)		
3.014	(move back and forth fast)		
3.015	(vibrations)		
3.016	(when something vibrates)		
3.017	(strong sound waves)		
3.018	(weak sound waves)		
3.019	(fast vibrations)		
3.020	(slow vibrations)		

Self Test 1

1.01   night

1.02   earth

1.03   rotation

1.04   twenty-four

1.05   model

1.06   axis

1.07   yes

1.08   no

1.09   no

1.010   yes

1.011   no

1.012   North Pole — top of the globe

1.013   axis — line through the earth

1.014   rotates — spins

1.015   revolve — go around the sun

1.016   globe — model of the earth

1.017   orbit — circle path

Self Test 2

2.01   365

2.02   12

2.03   24

2.04   4

2.05   1

2.06   rotate — spin

2.07   revolve — go around something

2.08   orbit — path

2.09   day — 24 hours

2.010   year — 365 days

2.011   The earth rotates and spins on its axis.

2.012   revolves or travels around the sun

2.013   at the same time

2.014   time zones

2.015   yes

2.016   no

2.017   yes

2.018   yes

2.019   yes

2.020   yes

Self Test 3

3.01   spring, summer, fall, and winter

3.02   Eastern, Central, Mountain, Pacific

3.03   the earth revolving around the sun on a tilted axis.

3.04   twenty-four hours

3.05   the earth rotating one time

3.06   a.   revolves

3.07   b   tilts

3.08   a.   summer

3.09   c.   fall

3.010   b.   spring

3.011   a.   summer

3.012   b.   fall

3.013   c.   winter

3.014   d.   spring

3.015   axis — the earth spins on it

3.016   rotates — spins

3.017   revolves — goes around

3.018   calendar — tells months and days

3.019   orbit — a path

3.020   time zone — Eastern Time Zone

3.021   colder — axis points away from sun

3.022   North Pole — top of axis

3.023   Big Dipper — made of stars

3.024   hotter — axis points toward sun

Self Test 1

1.01	hard	1.016	yes
1.02	God	1.017	no
1.03	should	1.018	no
1.04	boulder	1.019	yes
1.05	rock	1.020	yes
1.06	geologists	1.021	no
1.07	crust	1.022	no
1.08	igneous rock	1.023	basalt, granite
1.09	minerals	1.024	limestone, sandstone
1.010	volcano	1.025	gneiss, coal

1.011 A rock hound — studies rocks for fun.

1.012 Metamorphic — means changed.

1.013 Pressure — helps form rock.

1.014 Granite — is a rock formed from a volcano.

1.015 Basalt — is a hard rock formed from magma.

Self Test 2

2.01	d	2.011	mantle
2.02	e	2.012	wind
2.03	g	2.013	magma
2.04	f	2.014	geologists
2.05	a	2.015	basalt
2.06	b	2.016	core
2.07	c	2.017	water
2.08	earth	2.018	sedimentary
2.09	igneous	2.019	power
2.010	biologist	2.020	metamorphic

2.021	water	2.023	heat (cold)
2.022	wind	2.024	plants

Self Test 3

3.01	God	3.026	a lot of money
3.02	change	3.027	people who study rocks for fun
3.03	basalt	3.028	a. rocks
3.04	rocks		b. moon
3.05	Bible	3.029	a. statues
3.06	parents		b. marble
3.07	slates	3.030	studies rocks

3.08   Either order:
   a. plants
   b. animals

3.09   minerals

3.010   magma

3.011   d

3.012   f

3.013   e

3.014   a

3.015   c

3.016   g

3.017   b

3.018   no

3.019   no

3.020   yes

3.021   yes

3.022   yes

3.023   yes

3.024   no

3.025   Either order:
   a. build
   b. enjoy

Self Test 1

1.01	friction	1.018	Either order: a. wood b. coal
1.02	absorbed		
1.03	fire	1.019	Either order: a. oil b. gas
1.04	static		
1.05	fuel and oxygen		
1.06	radiant energy	1.020	natural gas
1.07	yes	1.021	d
1.08	yes	1.022	a
1.09	no	1.023	b
1.010	no	1.024	c
1.011	yes	1.025	e
1.012	yes	1.026	Any order: a. friction b. fire c. electricity d. your body e. the sun
1.013	no		
1.014	yes		
1.015	yes		
1.016	no		
1.017	yes		

Self Test 2

2.01	conduction	2.09	molecules
2.02	molecules	2.010	yes
2.03	expands	2.011	no
2.04	vibrate	2.012	no
2.05	friction	2.013	no
2.06	evaporation	2.014	yes
2.07	(friction) (your body) (the sun) (electricity) (fire)	2.015	yes
		2.016	oxygen
		2.017	vibrate
		2.018	radiation
		2.019	water
2.08	conduction X convection X radiation X	2.020	evaporated
		2.021	gold

| 2.022 | solid | 2.024 | a microscope |
| 2.023 | electricity | | |

Self Test 3

3.01	freeze	3.016	evaporates
3.02	sparks	3.017	expand
3.03	solar	3.018	move
3.04	benefits	3.019	conduct
3.05	problems	3.020	static
3.06	ecology	3.021	molecules
3.07	fire	3.022	yes
3.08	water pollution	3.023	no
3.09	air pollution	3.024	yes
3.010 -- 3.014	Any order:	3.025	yes
3.010	fire	3.026	no
3.011	friction	3.027	yes
3.012	electricity	3.028	no
3.013	our bodies	3.029	yes
3.014	the sun	3.030	yes
3.015	food		

Self Test 1

1.01   a.   C FRUITSANDVEGETABLES Z
        b.   DXFJL MILK OOILSANDFATS
        c.   BREADANDCEREAL IMSTWXR
        d.   MJMTH MEATANDEGGS IMDVY

1.02   Any order:
        a.   fruits
        b.   milk
        c.   bread and cereal
        d.   meat and eggs
        e.   vegetables
        f.   fats, oils, and sweets

1.03 — 1.04 Either order:

1.03   conscience

1.04   spirit

1.05   creative

1.06   teacher check

1.07   Any order:
        a.   water
        b.   correct temperature
        c.   minerals
        d.   sunlight
        e.   carbon dioxide

1.08   a.   egg
        b.   larva
        c.   pupa
        d.   adult

1.09   V

1.010   I

1.011   V

1.012   I

1.013   V

1.014   V

1.015   I

1.016   I

1.017   V

1.018   I

1.019   Any order:
        a.   air
        b.   food
        c.   exercise
        d.   rest

1.020   a, c

Self Test 2

2.01   Example:
        Matter is anything that has weight
        and takes up space.

2.02   Any order:
        a.   solid
        b.   liquid
        c.   gas

2.03   d

2.04   c

2.05   a

2.06   f

2.07   g

2.08   b

2.09   e

2.010   h

2.011   a.   summer   c.   winter
        b.   fall   d.   spring

2.012   hour

2.013   Pacific

2.014   magma

2.015   vertebrate

2.016   Either order:
   a.   heat
   b.   pressure

2.017   pupa

2.018   moth

2.019   Any order:
   a.   air
   b.   food
   c.   exercise
   d.   rest

2.020   a.   heat
   b.   wind
   c.   water
   d.   cold
   e.   ice
   f.   plants

Self Test 3

3.01   true

3.02   false

3.03   false

3.04   true

3.05   true

3.06   Heat is the energy of vibrating molecules.

3.07   Any order:
   a.   friction
   b.   fire
   c.   electricity
   d.   your body
   e.   sun

3.08   oxygen

3.09   creative

3.010   chlorophyll

3.011   sunlight

3.012   igneous

3.013   solid

3.014   gas

3.015   liquid

3.016   liquid

3.017   solid

3.018   gas

3.019   twenty-four

3.020   a day

3.021   365

3.022   seasons

3.023   four

Science 301
LIFEPAC Test

1. in mouth
2. openings in nose
3. breathe out
4. breathe in
5. carries messages
6. man has
7. 3
8. 5
9. 1
10. 4
11. 2
12. exercise
13. oxygen
14. carbon dioxide
15. Joseph Priestley

16. pulse
17. Example: People have consciences.
18. Example: People have spirits.
19. Example: Animals and people can breathe.
20. Example: Animals and people need air, water, and food.
21. Example: Animals and people digest food.
22. body
23. pulse
24. hairs
25. faster
26. diaphragm
27. rest

Science 302
LIFEPAC Test

1. how hot or cold it is
2. green material in plants
3. to put a name on something
4. a plant just beginning to grow
5. a tiny plant with its food
6. God
7. stores
8. oxygen
9. stem
10. magnifying glass
11. degrees
12. soil
13. fruit

14. seed
15. roots
16. leaves
17. stems
18. take in water and minerals
19. take water and minerals to the leaves
20. give off oxygen and water; make food
21. water
22. minerals
23. Carbon dioxide
24. sunlight
25. right temperature

Science 303
LIFEPAC Test

1. temperature
2. expands
3. mammals
4. freeze, ice
5. vertebrates
6. birds, mammals
7. lungs
8. metamorphosis
9. potato bugs
10. Soil, water

11. Any order: fish, amphibians, reptiles, birds, mammals
12. Any order: temperature, water, light, soil
13. egg, larva, pupa, adult
14. a. 212° F   b. 100° C
    c. 32° F   d. 0° C
15. 32° F
16. a. antenna
    b. head
    c. thorax
    d. abdomen

Science 304
LIFEPAC Test

1. junk
2. good
3. yes
4. yes
5. yes
6. yes
7. no
8. yes
9. no
10. yes
11. yes

Any order:
12. fruits: apple
13. milk: milk
14. meat & eggs: lamb chops
15. bread & cereal: corn flakes

16. vegetables: corn
17. fats, oils, sweets: candy
18. grow
19. shoulder
20. clean
21. water
22. clothes

Examples:
23. beef/steaks   meat & eggs
24. potato   vegetables
25. salad   vegetables
26. bread   bread & cereal
27. milk   milk
28. ice cream   milk

Science 305
LIFEPAC Test

1. changes shape and size
2. has size and changing shape
3. fixed shape and size
4. how much matter is in something
5. a person that studies chemistry
6. space used by matter
7. cannot be seen
8. melting ice cream
9. tiny bits of matter
10. a forest fire

11. no
12. yes
13. yes
14. yes
15. no
16. solid
17. matter
18. physical change
19. chemical change
20. physical change

Science 306
LIFEPAC Test

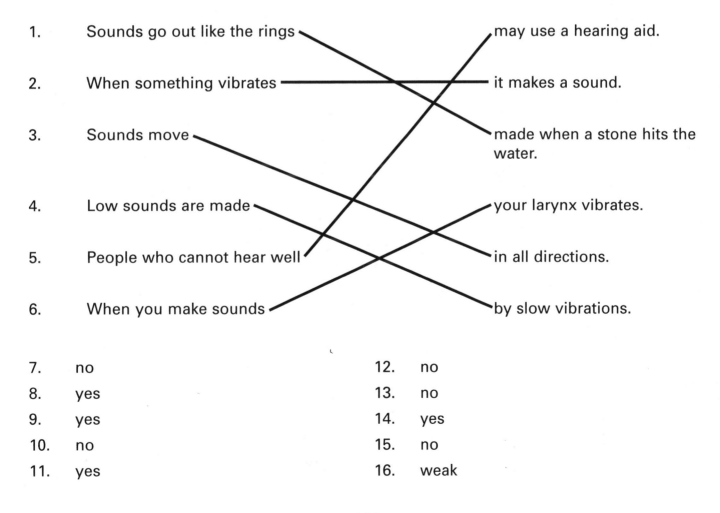

1. Sounds go out like the rings — made when a stone hits the water.
2. When something vibrates — it makes a sound.
3. Sounds move — in all directions.
4. Low sounds are made — by slow vibrations.
5. People who cannot hear well — may use a hearing aid.
6. When you make sounds — your larynx vibrates.

7. no
8. yes
9. yes
10. no
11. yes

12. no
13. no
14. yes
15. no
16. weak

17. sound
18. vibrate
19. ears
20. loud
21. animals, people, and things

22. eardrum
23. neck
24. direction
25. louder

Science 307
LIFEPAC Test

1. once
2. revolves
3. night
4. day
5. sun
6. North Pole
7. summer
8. zones
9. calendar
10. rotations
11. no
12. no
13. yes
14. yes
15. yes
16. yes

17. no
18. yes
19. yes
20. no
21. Any order
   a. summer
   b. fall
   c. winter
   d. spring
22. Any order
   a. Eastern
   b. Central
   c. Mountain
   d. Pacific
23. time
24. seasons
25. 365
26. 24

Science 308
LIFEPAC Test

1. b
2. d
3. e
4. g
5. a
6. j
7. f

8. c
9. i
10. h
11. yes
12. no
13. no
14. yes

15. yes

16. volcano

17. magma

18. Either order:
    a. heat
    b. pressure

19. Either order:
    a. water
    b. wind

20. minerals

21. rock hounds

22. gems

23. plants

24. igneous

25. sedimentary

26. metamorphic

Science 309
LIFEPAC Test

1. friction

2. fire

3. absorbed

4. fuel

5. resources

6. conduction

7. expand

8. molecules

9. pollution

10. conduction
    convection
    radiation

11. fire
    electricity
    friction
    the sun
    your body

12. rise   X
    expand   X

13. benefits

14. problem

15. environment

16. ecology

17. expand

18. contract

19. God-given

20. man-made

Science 310
LIFEPAC Test

1. Any order
    a. water
    b. correct temperature
    c. minerals
    d. sunlight
    e. carbon dioxide

2. Any order:
    a. air
    b. food
    c. exercise
    d. rest

3. Any order:
    a. solid
    b. liquid
    c. gas

4. g

5. f

6. d

7. c

8. b

9. e

10. a

11. h

12. i

13. the energy of vibrating molecules

14. true

15. false

16. true

17. false

18. true

19.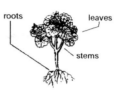

20. Any order:
    a. friction
    b. fire
    c. electricity
    d. your body
    e. sun